Manual on Presentation of Data and Control Chart Analysis

7th Edition

Prepared by

COMMITTEE E-11 ON QUALITY AND STATISTICS

Stock #: **MNL7A**

Revision of Special Technical Publication (STP) 15D

ASTM International ▪ 100 Barr Harbor Drive ▪ PO Box C700
West Conshohocken, PA 19428-2959

Library of Congress Cataloging-in-Publication Data

Manual on presentation of data and control chart analysis / prepared
by the Committee E-11 on statistical control.
(ASTM manual series ; MNL 7)
Includes bibliographical references.
ISBN 0-8031-1289-0
1. Materials—Testing—Handbooks, manuals, etc. 2. Quality
control—Statistical methods—Handbooks, manuals, etc. I. ASTM
Committee E-11 on Statistical Methods. II. Series.
TA410.M355 1989
620.1′1′0287—dc20   89-18047
                                                              CIP

Printed in Bridgeport, NJ
February 2002

Foreword

THIS ASTM Manual on Presentation of Data and Control Chart Analysis is the sixth revision of the original ASTM Manual on Presentation of Data first published in 1933. This sixth revision was prepared by the ASTM E11.10 Subcommittee on Sampling and Data Analysis, which serves the ASTM Committee E-11 on Quality and Statistics.

Contents

PREFACE

THIS *Manual on the Presentation of Data and Control Chart Analysis* (MNL 7), was prepared by ASTM's Committee E-11 on Quality and Statistics to make available to the ASTM INTERNATIONAL membership, and others, information regarding statistical and quality control methods, and to make recommendations for their application in the engineering work of the Society. The quality control methods considered herein are those methods that have been developed on a statistical basis to control the quality of product through the proper relation of specification, production, and inspection as parts of a continuing process.

The purposes for which the Society was founded—the promotion of knowledge of the materials of engineering, and the standardization of specifications and the methods of testing—involve at every turn the collection, analysis, interpretation, and presentation of quantitative data. Such data form an important part of the source material used in arriving at new knowledge and in selecting standards of quality and methods of testing that are adequate, satisfactory, and economic, from the standpoints of the producer and the consumer.

Broadly, the three general objects of gathering engineering data are to discover: (1) physical constants and frequency distributions, (2) the relationships—both functional and statistical—between two or more variables, and (3) causes of observed phenomena. Under these general headings, the following more specific objectives in the work of ASTM International may be cited: (a) to discover the distributions of quality characteristics of materials which serve as a basis for setting economic standards of quality, for comparing the relative merits of two or more materials for a particular use, for controlling quality at desired levels, for

predicting what variations in quality may be expected in subsequently produced material; to discover the distributions of the errors of measurement for particular test methods, which serve as a basis for comparing the relative merits of two or more methods of testing, for specifying the precision and accuracy of standard tests, for setting up economical testing and sampling procedures; (b) to discover the relationship between two or more properties of a material, such as density and tensile strength; and (c) to discover physical causes of the behavior of materials under particular service conditions; to discover the causes of nonconformance with specified standards in order to make possible the elimination of assignable causes and the attainment of economic control of quality.

Problems falling in these categories can be treated advantageously by the application of statistical methods and quality control methods. This Manual limits itself to several of the items mentioned under (a). **PART 1** discusses frequency distributions, simple statistical measures, and the presentation, in concise form, of the essential information contained in a single set of n observations. **PART 2** discusses the problem of expressing ± limits of uncertainty for various statistical measures, together with some working rules for rounding-off observed results to an appropriate number of significant figures. **PART 3** discusses the control chart method for the analysis of observational data obtained from a series of samples, and for detecting lack of statistical control of quality.

The present Manual is the sixth revision of earlier work on the subject. The original *ASTM Manual on Presentation of Data, STP 15*, issued in 1933 was prepared by a special committee of former

Subcommittee IX on Interpretation and Presentation of Data of ASTM Committee E-1 on Methods of Testing. In 1935, Supplement A on Presenting ± Limits of Uncertainty of an Observed Average and Supplement B on "Control Chart" Method of Analysis and Presentation of Data were issued. These were combined with the original manual and the whole, with minor modifications, was issued as a single volume in 1937. The personnel of the Manual Committee that undertook this early work were: H. F. Dodge, W. C. Chancellor, J. T. McKenzie, R. F. Passano, H. G. Romig, R. T. Webster, and A. E. R. Westman. They were aided in their work by the ready cooperation of the Joint Committee on the Development of Applications of Statistics in Engineering and Manufacturing (sponsored by ASTM International and the American Society of Mechanical Engineers (ASME)) and especially of the chairman of the Joint Committee, W. A. Shewhart. The nomenclature and symbolism used in this early work were adopted in 1941 and 1942 in the American War Standards on Quality Control (Z1.1, Z1.2, and Z1.3) of the American Standards Association, and its Supplement B was reproduced as an appendix with one of these standards.

In 1946, ASTM Technical Committee E-11 on Quality Control of Materials was established under the chairmanship of H. F. Dodge, and the manual became its responsibility. A major revision was issued in 1951 as *ASTM Manual on Quality Control of Materials, STP 15C*. The Task Group that undertook the revision of **PART 1** consisted of R. F. Passano, Chairman, H. F. Dodge, A. C. Holman, and J. T. McKenzie. The same task group also revised **PART 2** (the old Supplement A) and the task group for revision of **PART 3** (the old Supplement B) consisted of A. E. R. Westman, Chairman, H. F. Dodge, A. I. Peterson, H. G. Romig, and L. E. Simon. In this 1951 revision, the term "confidence limits" was introduced and constants for computing 0.95 confidence limits were added to the constants for 0.90 and 0.99 confidence limits presented in prior printings. Separate treatment was given to control charts for "number of defectives," "number of defects," and "number of defects per unit" and material on control charts for individuals was added. In subsequent editions, the term "defective" has been replaced by "nonconforming unit" and "defect" by "nonconformity" to agree with definitions adopted by the American Society for Quality Control in 1978 (See the American National Standard, ANSI/ASQC A1-1987, *Definitions, Symbols, Formulas and Tables for Control Charts*.)

There were more printings of *ASTM STP 15C*, one in 1956 and a second in 1960. The first added the ASTM Recommended Practice for Choice of Sample Size to Estimate the Average Quality of a Lot or Process (E 122) as an Appendix. This recommended practice had been prepared by a task group of ASTM Committee E-11 consisting of A. G. Scroggie, Chairman, C. A. Bicking, W. E. Deming, H. F. Dodge, and S. B. Littauer. This Appendix was removed from that edition because it is revised more often than the main text of this Manual. The current version of E 122, as well as of other relevant ASTM International publications, may be procured from ASTM International. (See the list of references at the back of this Manual.)

In the 1960 printing, a number of minor modifications were made by an ad hoc committee consisting of Harold Dodge, Chairman, Simon Collier, R. H. Ede, R. J. Hader, and E. G. Olds.

The principal change in *ASTM STP 15C* introduced in *ASTM STP 15D* was the redefinition of the sample standard deviation to be $s = \sqrt{\sum (X_i - \overline{X})^2 \Big/ (n-1)}$. This change required numerous changes throughout the Manual in mathematical equations and formulas, tables, and numerical illustrations. It also led to a sharpening of distinctions between sample values, universe values, and standard values that were not formerly deemed necessary.

New material added in *ASTM STP 15D* included the following items. The sample measure of kurtosis, g_2, was introduced. This addition led to a revision of Table 8 and Section 34 of **PART 1**. In **PART 2**, a brief discussion of the determination of confidence limits for a universe standard deviation and a universe proportion was included. The Task Group responsible for this fourth revision of the Manual consisted of A. J. Duncan, Chairman R. A. Freund, F. E. Grubbs, and D. C. McCune.

In the twenty-two years between the appearance of *ASTM STP 15D* and *Manual on Presentation of Data and Control Chart Analysis, 6th Edition* there were two reprintings without significant changes. In that period a number of misprints and minor inconsistencies were found in *ASTM STP 15D*. Among these were a few erroneous calculated values of control chart factors appearing in tables of **PART 3.** While all of these errors were small, the mere fact that they existed suggested a need to recalculate all tabled control chart factors. This task was carried out by A. T. A. Holden, a student at the Center for Quality and Applied Statistics at the Rochester Institute of Technology, under the general guidance of Professor E. G. Schilling of Committee E 11. The tabled values of control chart factors have been corrected where found in error. In addition, some ambiguities and inconsistencies between the text and the examples on attribute control charts have received attention.

A few changes were made to bring the Manual into better agreement with contemporary statistical notation and usage. The symbol μ (Greek "mu") has replaced X' (and $\overline{X'}$) for the universe average of measurements (and of sample averages of those measurements.) At the same time, the symbol σ has replaced σ' as the universe value of standard deviation. This entailed replacing σ by $s_{(rms)}$ to denote the sample root-mean-square deviation. Replacing the universe values p', u' and c' by Greek letters was thought worse than leaving them as they are. Section 33, **PART 1**, on distributional information conveyed by Chebyshev's inequality, has been revised.

Summary of changes in definitions and notations.

MNL 7	STP 15D
μ, σ, p', u', c'	$\overline{X'}$, σ', p', u', c'
(= universe values)	(= universe values)
μ_0, σ_0, u_0, c_0	$\overline{X'}_0$, σ_0', p_0', u_0', c_0'
(= standard values)	(= standard values)

In the twelve-year period since this Manual was revised again, three developments were made that had an increasing impact on the presentation of data and control chart analysis. The first was the introduction of a variety of new tools of data analysis and presentation. The effect to date of these developments is not fully reflected in **PART 1** of this edition of the Manual, but an example of the "stem and leaf" diagram is now presented in Section 15. *Manual on Presentation of Data and Control Chart Analysis, 6th Edition* from the first has embraced the idea that the control chart is an all-important tool for data analysis and presentation. To integrate properly the discussion of this established tool with the newer ones presents a challenge beyond the scope of this revision.

The second development of recent years strongly affecting the presentation of data and control chart analysis is the greatly increased capacity, speed, and availability of personal computers and sophisticated hand calculators. The computer revolution has not only enhanced capabilities for data analysis and presentation, but has enabled techniques of high speed real-time data-taking, analysis, and process control, which years ago would have been unfeasible, if not unthinkable. This has made it desirable to include some discussion of practical approximations for control chart factors for rapid if not real-time application. Supplement A has been considerably revised as a result. (The issue of approximations was raised by Professor A. L. Sweet of Purdue University.) The approximations presented in this Manual presume the computational ability to take squares and square roots of rational numbers without using tables. Accordingly, the Table of Squares and Square Roots that appeared as an Appendix to ASTM STP 15D was removed from the previous revision. Further discussion of approximations appears in Notes 8 and 9 of Supplement B, **PART 3**. Some of the approximations presented in **PART 3** appear to be new and assume mathematical forms suggested in part by unpublished work of Dr. D. L. Jagerman of AT&T Bell Laboratories on the ratio of gamma functions with near arguments.

The third development has been the refinement of alternative forms of the control chart, especially the exponentially weighted moving average chart and the cumulative sum ("cusum") chart. Unfortunately, time was lacking to include discussion of these developments in the fifth revision, although references are given. The assistance of S. J Amster of AT&T Bell Laboratories in providing recent references to these developments is gratefully acknowledged.

Manual on Presentation of Data and Control Chart Analysis, 6th Edition by Committee E-11 was initiated by M. G. Natrella with the help of comments from A. Bloomberg, J. T. Bygott, B. A. Drew, R. A. Freund, E. H. Jebe, B. H. Levine, D. C. McCune, R. C. Paule, R. F. Potthoff, E. G. Schilling and R. R. Stone. The revision was completed by R. B. Murphy and R. R. Stone with further comments from A. J. Duncan, R. A. Freund, J. H. Hooper, E. H. Jebe and T. D. Murphy.

Manual on Presentation of Data and Control Chart Analysis, 7th Edition has been directed at bringing the discussions around the various methods covered in **PART 1** up to date. Especially, in the areas of whole number frequency distributions, empirical percentiles, and order statistics. As an example, an extension of the stem-and-leaf diagram has been added which is termed an "ordered stem-and-leaf," which makes it easier to locate the quartiles of the distribution. These quartiles, along with the maximum and minimum values, are then used in the construction of a box plot.

In **PART 3**, additional material has been included to discuss the idea of risk, namely, the alpha (α) and beta (β) risks involved in the decision-making process based on data; and tests for assessing evidence of nonrandom behavior in process control charts.

Also, use of the $s_{(rms)}$ statistic has been minimized in this revision in favor of the sample standard deviation s to reduce confusion as to their use. Furthermore, the graphics and tables throughout the text have been repositioned so that they appear more closely to their discussion in the text.

Manual on Presentation of Data and Control Chart Analysis, 7th Edition by Committee E-11 was initiated and led by Dean V. Neubauer, Chairman of the E11.10 Subcommittee on Sampling and Data Analysis that oversees this document. Additional comments from Steve Luko, Charles Proctor, Paul Selden, Greg Gould, Frank Sinibaldi, Ray Mignogna, Neil Ullman, Thomas D. Murphy, and R. B. Murphy were instrumental in the vast majority of the revisions made in this sixth revision. Thanks must also be given to Kathy Dernoga and Monica Siperko of the ASTM International New Publications department for their efforts in the publication of this edition.

1

Presentation of Data

PART 1 IS CONCERNED solely with presenting information about a given sample of data. It contains no discussion of inferences that might be made about the population from which the sample came.

SUMMARY

Bearing in mind that no rules can be laid down to which no exceptions can be found the committee believes that if the recommendations presented are followed, the presentations will contain the essential information for a majority of the uses made of ASTM data.

RECOMMENDATIONS FOR PRESENTATION OF DATA

Given a sample of n observations of a single variable obtained under the same essential conditions:

1. Present as a minimum, the average, the standard deviation, and the number of observations. *Always* state the number of observations.

2. Also, present the values of the maximum and minimum observations. Any collection of observations may contain mistakes. If errors occur in the collection of the data, then correct the data values, but do <u>not</u> discard or change any other observations.

3. The average and standard deviation are sufficient to describe the data, particularly so when they follow a Normal distribution.

To see how the data may depart from a Normal distribution, prepare the grouped frequency distribution and its histogram. Also, calculate skewness, g_1, and kurtosis, g_2.

4. If the data seem not to be normally distributed, then one should consider presenting the median and percentiles (discussed in Section 6), or consider a transformation to make the distribution more normally distributed. The advice of a statistician should be sought to help determine which, if any, transformation is appropriate to suit the user's needs.

5. Present as much evidence as possible that the data were obtained under controlled conditions.

6. Present relevant information on precisely (*a*) the field of application within which the measurements are believed valid and (*b*) the conditions under which they were made.

GLOSSARY OF SYMBOLS USED IN PART 1

f *Observed frequency* (number of observations) in a single bin of a frequency distribution

g_1 *Sample coefficient of skewness*, a measure of skewness, or lopsidedness of a distribution

g_2 *Sample coefficient of kurtosis*

n Number of observed values (observations)

p *Sample relative frequency* or *proportion*, the ratio of the number of occurrences of a given type to the total possible number of occurrences, the ratio of the number of observations in any stated interval to the total number of observations; *sample fraction nonconforming* for measured values the ratio of the number of observations lying outside specified limits (or beyond a specified limit) to the total number of observations

R *Sample range*, the difference between the largest observed value and the smallest observed value.

s *Sample standard deviation*

s^2 *Sample variance*

cv *Sample coefficient of variation*, a measure of relative dispersion based on the standard deviation (see Sect. 31)

X Observed values of a measurable characteristic; specific observed values are designated X_1, X_2, X_3, etc. in order of measurement, and $X_{(1)}$, $X_{(2)}$, $X_{(3)}$, etc. in order of their size, where $X_{(1)}$ is the smallest or minimum observation and $X_{(n)}$ is the largest or maximum observation in a sample of observations; also used to designate a measurable characteristic

$\overline{X}$ *Sample average* or *sample mean*, the sum of the n observed values in a sample divided by n

NOTE

The sample proportion p is an example of a sample average in which each observation is either a 1, the occurrence of a given type, or a 0, the nonoccurrence of the same type. The sample average is then exactly the ratio, *p*, of the total number of occurrences to the total number possible in the sample, *n*.

If reference is to be made to the population from which a given sample came, the following symbols should be used.

γ_1 *Population skewness* defined as the expected value (see Note) of $(X - \mu)^3$ divided by σ^3. It is spelled and pronounced "gamma one."

γ_2 *Population coefficient of kurtosis* defined as the amount by which the expected value (see Note) of $(X - \mu)^4$ divided by σ^4 exceeds or falls short of 3; it is spelled and pronounced "gamma two."

μ *Population average* or *universe mean* defined as the expected value (see Note) of X; thus $E(X) = \mu$, spelled "mu" and pronounced "mew."

p' *Population relative frequency*

σ *Population standard deviation*, spelled and pronounced "sigma."

σ^2 *Population variance* defined as the expected value (see Note) of the square of a deviation from the universe mean; thus $E[(X - \mu)^2] = \sigma^2$

CV *Population coefficient of variation* defined as the population standard deviation divided by the population mean, also called the *relative standard deviation*, or *relative error*. (see Sect. 31)

NOTE

If a set of data is homogeneous in the sense of Section 3 of **PART 1**, it is usually safe to apply statistical theory and its concepts, like that of an *expected value*, to the data to assist in its analysis and interpretation. Only then is it meaningful to speak of a population average or other characteristic relating to a population (relative) frequency distribution function of *X*. This function commonly assumes the form of *f(x)*, which is the probability (relative frequency) of an observation having exactly the value *X*, or the form of *f(x)dx*, which is the probability an observation has a value between *x* and *x + dx*. Mathematically the *expected value* of a function of *X*, say *h(X)*, is defined as the sum (for discrete data) or integral (for continuous data) of that function times the probability of *X* and written *E[h(X)]*. For example, if the probability of *X* lying between *x* and *x + dx* based on continuous data is *f(x)dx*, then the expected value is

$$\int h(x)f(x)dx = E[h(x)].$$

If the probability of *X* lying between *x* and *x + dx* based on continuous data is *f(x)dx*, then the expected value is

$$\Sigma h(x)f(x)dx = E[h(x)].$$

Sample statistics, like $\overline{X}$, s^2, g_1, and g_2, also have expected values in most practical cases, but these expected values relate to the population frequency distribution of *entire samples* of n observations each, rather than of individual observations. The expected value of $\overline{X}$ is μ, the same as that of an individual observation regardless of the population frequency distribution of X, and $E(s^2) = \sigma^2$ likewise, but $E(s)$ is less than σ in all cases and its value depends on the population distribution of X.

INTRODUCTION

1. Purpose

PART 1 of the Manual discusses the application of statistical methods to the problem of: (*a*) condensing the information contained in a sample of observations, and (*b*) presenting the essential information in a concise form more readily interpretable than the unorganized mass of original data.

Attention will be directed particularly to quantitative information on measurable characteristics of materials and manufactured products. Such characteristics will be termed *quality characteristics*.

2. Type of Data Considered

Consideration will be given to the treatment of a sample of *n* observations of a single variable. Figure 1 illustrates two general types: (*a*) the first type is a series of *n* observations representing single measurements of the same quality characteristic of *n* similar things, and (*b*) the second type is a series of *n* observations representing *n* measurements of the same quality characteristic of one thing.

The observations in Figure 1 are denoted as X_i, where $i = 1, 2, 3, \ldots, n$. Generally, the subscript will represent the time sequence in which the observations were taken from a process or measurement. In this sense, we may consider the order of the data in Table 1 as being represented in a time-ordered manner.

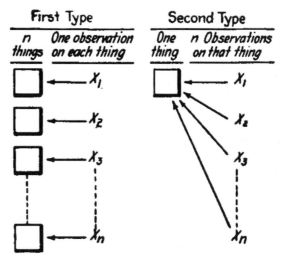

FIG. 1—Two general types of data.

Data of the first type are commonly gathered to furnish information regarding the *distribution* of the quality of the material itself, having in mind possibly some more specific purpose; such as the establishment of a quality standard or the determination of conformance with a specified quality standard, for example, 100 observations of transverse strength on 100 bricks of a given brand.

Data of the second type are commonly gathered to furnish information regarding the errors of measurement for a particular test method, for example, 50-micrometer measurements of the thickness of a test block.

NOTE

The quality of a material in respect to some particular characteristic, such as tensile strength, is better represented by a frequency distribution function, than by a single-valued constant.

The variability in a group of observed values of such a quality characteristic is made up of two parts: variability of the material itself, and the errors of measurement. In some practical problems, the error of measurement may be large compared with the variability of the material; in others, the converse may be true. In any case, if one is interested in discovering the objective frequency distribution of the quality of the material, consideration must be given to correcting the errors of measurement (This is discussed in Ref. **1**, pp. 379-384, in the seminal book on control chart methodology by Walter A. Shewhart.).

3. Homogeneous Data

While the methods here given may be used to condense any set of observations, the results obtained by using them may be of little value from the standpoint of interpretation unless the data are good in the first place and satisfy certain requirements.

To be useful for inductive generalization, any sample of observations that is treated as a single group for presentation purposes should represent a series of measurements, all made under essentially the same test conditions, on a material or product, all of which has been produced under essentially the same conditions.

If a given sample of data consists of two or more subportions collected under different test conditions or representing material produced under different conditions, it should be considered as two or more separate subgroups of observations, each to be treated independently in the analysis. Merging of such subgroups, representing significantly different conditions, may lead to a condensed presentation that will be of little practical value. Briefly, any sample of observations to which these methods are applied should be *homogeneous*.

In the illustrative examples of **PART 1**, each sample of observations will be assumed to be homogeneous, that is, observations from a common universe of causes. The analysis and presentation by control chart methods of data obtained from several samples or capable of subdivision into subgroups on the basis of relevant engineering information is discussed in **PART 3** of this Manual. Such methods enable one to determine whether for practical

TABLE 1. Three groups of original data.

(a) Transverse Strength of 270 Bricks of a Typical Brand, psi[a]

860	1320	820	1040	1000	1010	1190	1180	1080	1100	1130
920	1100	1250	1480	1150	740	1080	860	1000	810	1000
1200	830	1100	890	270	1070	830	1380	960	1360	730
850	920	940	1310	1330	1020	1390	830	820	980	1330
920	1070	1630	670	1150	1170	920	1120	1170	1160	1090
1090	700	910	1170	800	960	1020	1090	2010	890	930
830	880	870	1340	840	1180	740	880	790	1100	1260
1040	1080	1040	980	1240	800	860	1010	1130	970	1140
1510	1060	840	940	1110	1240	1290	870	1260	1050	900
740	1230	1020	1060	990	1020	820	1030	860	850	890
1150	860	1100	840	1060	1030	990	1100	1080	1070	970
1000	720	800	1170	970	690	1020	890	700	880	1150
1140	1080	990	570	790	1070	820	580	820	1060	980
1030	960	870	800	1040	820	1180	1350	1180	950	1110
700	860	660	1180	780	1230	950	900	760	1380	900
920	1100	1080	980	760	830	1220	1100	1090	1380	1270
860	990	890	940	910	1100	1020	1380	1010	1030	950
950	880	970	1000	990	830	850	630	710	900	890
1020	750	1070	920	870	1010	1230	780	1000	1150	1360
1300	970	800	650	1180	860	1150	1400	880	730	910
890	1030	1060	1610	1190	1400	850	1010	1010	1240	
1080	970	960	1180	1050	920	1110	780	780	1190	
910	1100	870	980	730	800	800	1140	940	980	
870	970	910	830	1030	1050	710	890	1010	1120	
810	1070	1100	460	860	1070	880	1240	940	860	

purposes a given sample of observations may be considered to be homogeneous.

4. Typical Examples of Physical Data

Table 1 gives three typical sets of observations, each one of these datasets represents measurements on a sample of units or specimens selected in a random manner to provide information about the quality of a larger quantity of material—the general output of one brand of brick, a production lot of galvanized iron sheets, and a shipment of hard drawn copper wire. Consideration will be given to ways of arranging and condensing these data into a form better adapted for practical use.

TABLE 1. Continued.

(b) Weight of Coating of 100 Sheets of Galvanized Iron Sheets, oz/ft^2 [b]					(c) Breaking Strength of Ten Specimens of 0.104-in. Hard-Drawn Copper Wire, lb[c]
1.467	1.603	1.577	1.563	1.437	578
1.623	1.603	1.577	1.393	1.350	572
1.520	1.383	1.323	1.647	1.530	570
1.767	1.730	1.620	1.620	1.383	568
1.550	1.700	1.473	1.530	1.457	572
1.533	1.600	1.420	1.470	1.443	570
1.377	1.603	1.450	1.337	1.473	570
1.373	1.477	1.337	1.580	1.433	572
1.637	1.513	1.440	1.493	1.637	576
1.460	1.533	1.557	1.563	1.500	584
1.627	1.593	1.480	1.543	1.607	
1.537	1.503	1.477	1.567	1.423	
1.533	1.600	1.550	1.670	1.573	
1.337	1.543	1.637	1.473	1.753	
1.603	1.567	1.570	1.633	1.467	
1.373	1.490	1.617	1.763	1.563	
1.457	1.550	1.477	1.573	1.503	
1.660	1.577	1.750	1.537	1.550	
1.323	1.483	1.497	1.420	1.647	
1.647	1.600	1.717	1.513	1.690	

[a] Measured to the nearest 10 psi. Test method used was ASTM Method of Testing Brick and Structural Clay (C 67). Data from *ASTM Manual for Interpretation of Refractory Test Data*, 1935, p. 83.

[b] Measured to the nearest 0.01 oz/ft^2 of sheet, averaged for three spots. Test method used was ASTM Triple Spot Test of Standard Specifications for Zinc-Coated (Galvanized) Iron or Steel Sheets (A 93). This has been discontinued and was replaced by ASTM Specification for General Requirements for Steel Sheet, Zinc-Coated (Galvanized) by the Hot-Dip Process (A 525). Data from laboratory tests.

[c] Measured to the nearest 2 lb. Test method used was ASTM Specification for Hard-Drawn Copper Wire (B 1). Data from inspection report.

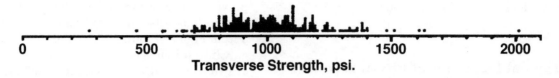

Fig. 2—Showing graphically the ungrouped frequency distribution of a set of observations. Each dot represents one brick, data of Table 2(a).

UNGROUPED WHOLE NUMBER DISTRIBUTION

5. Ungrouped Distribution

An arrangement of the observed values in ascending order of magnitude will be referred to in the Manual as the *ungrouped frequency distribution* of the data, to distinguish it from the grouped frequency distribution defined in Section 8. A further adjustment in the scale of the ungrouped distribution produces the whole number distribution. For example, the data of Table 1(a) were multiplied by 10^{-1}, and those of Table 1(b) by 10^3, while those of Table 1(c) were already whole numbers. If the data carry digits past the decimal point, just round until a tie (one observation equals some other) appears and then scale to whole numbers. Table 2 presents ungrouped frequency distributions for the three sets of observations given in Table 1.

Figure 2 shows graphically the ungrouped frequency distribution of Table 2(*a*). In the graph, there is a minor grouping in terms of the unit of measurement. For the data of Fig. 2, it is the "rounding-off" unit of 10 psi. It is rarely desirable to present data in the manner of Table 1 or Table 2. The mind cannot grasp in its entirety the meaning of so many numbers; furthermore, greater compactness is required for most of the practical uses that are made of data.

6. Empirical Percentiles and Order Statistics

As should be apparent, the ungrouped whole number distribution may differ from the original data by a scale factor (some power of ten), by some rounding and by having been sorted from smallest to largest. These features should make it easier to convert from an ungrouped to a grouped frequency distribution. More importantly, they allow calculation of the *order statistics* that will aid in finding ranges of the distribution wherein lie specified proportions of the observations. A collection of observations is often seen as only a sample from a potentially huge population of observations and one aim in studying the sample may be to say what proportions of values in the population lie in certain ranges. This is done by calculating the *percentiles* of the distribution. We will see there are a number of ways to do this but we begin by discussing order statistics and empirical estimates of percentiles.

A glance at Table 2 gives some information not readily observed in the original data set of Table 1. The data in Table 2 are arranged in increasing order of magnitude. When we arrange any data set like this the resulting ordered sequence of values are referred to as *order statistics*. Such ordered arrangements are often of value in the initial stages of an analysis. In this context, we use subscript notation and write $X_{(i)}$ to denote the ith order statistic. For a sample of n values the order statistics are $X_{(1)} \leq X_{(2)} \leq X_{(3)} \leq \ldots \leq X_{(n)}$. The index i is sometimes called the *rank* of the data point to which it is attached. For a sample size of n values, the first order statistic is the smallest or minimum value and has rank 1. We write this as $X_{(1)}$. The nth order statistic is the largest or maximum value and has rank n. We write this as $X_{(n)}$. The ith order statistic is written as $X_{(i)}$, for $1 \leq i \leq n$. For the breaking strength data in Table 2c, the order statistics are: $X_{(1)}=568$, $X_{(2)}=570$, ... , $X_{(10)}=584$.

When ranking the data values, we may find some that are the same. In this situation, we say that a matched set of values constitutes a *tie*. The proper rank assigned to values that make up the tie is calculated by averaging the

TABLE 2. Ungrouped frequency distributions in tabular form.

(*a*) Transverse Strength, psi (data of Table 1 (*a*))

270	780	830	870	920	970	1020	1070	1100	1180	1310
460	780	830	880	920	980	1020	1070	1100	1180	1320
570	780	830	880	920	980	1020	1070	1100	1180	1330
580	790	840	880	920	980	1020	1070	1100	1180	1330
630	790	840	880	920	980	1020	1070	1110	1180	1340
650	800	840	880	930	980	1020	1070	1110	1180	1350
660	800	850	880	940	980	1020	1070	1110	1180	1360
670	800	850	890	940	990	1030	1080	1120	1190	1360
690	800	850	890	940	990	1030	1080	1120	1190	1380
700	800	850	890	940	990	1030	1080	1130	1190	1380
700	800	860	890	940	990	1030	1080	1130	1200	1380
700	800	860	890	950	990	1030	1080	1140	1220	1380
710	810	860	890	950	1000	1030	1080	1140	1230	1390
710	810	860	890	950	1000	1040	1080	1140	1230	1400
720	820	860	890	950	1000	1040	1090	1150	1230	1400
730	820	860	900	960	1000	1040	1090	1150	1240	1480
730	820	860	900	960	1000	1040	1090	1150	1240	1510
730	820	860	900	960	1000	1050	1090	1150	1240	1610
740	820	860	900	960	1010	1050	1100	1150	1240	1630
740	820	860	910	970	1010	1050	1100	1150	1250	2010
740	820	870	910	970	1010	1060	1100	1160	1260	
750	830	870	910	970	1010	1060	1100	1170	1260	
760	830	870	910	970	1010	1060	1100	1170	1270	
760	830	870	910	970	1010	1060	1100	1170	1290	
780	830	870	920	970	1010	1060	1100	1170	1300	

ranks that would have been determined by the procedure above in the case where each value was different from the others. For example, there are many ties present in Table 2. The rank associated with the three values of 700 would be the average of the ranks as if they were 700, 701, and 702, respectively. In other words, we see that the values of 700 occur in the 10th, 11th, and 12th positions, or represented as $X_{(10)}$, $X_{(11)}$, and $X_{(12)}$, respectively, if they were unequal. Thus, the value of 700 should carry a rank equal to $(10+11+12)/3 = 11$, and each value specified as $X_{(11)}$.

The order statistics can be used for a variety of purposes, but it is for estimating the percentiles that they are used here. A *percentile* is a value that divides a distribution to leave a given fraction of the observations less than that value. For example, the 50th percentile, typically referred to as the *median*, is a value such that half of the observations exceed it and half are below it. The 75th percentile is a value such that 25% of the observations exceed it and 75% are below it. The 90th percentile is a value such that 10% of the observations exceed it and 90% are below it.

To aid in understanding the formulas that follow, consider finding the percentile that best corresponds to a given order statistic. Although there are several answers to this question, one of the simplest is to realize that a sample of size *n* will partition the distribution from which it came into $n+1$ compartments as illustrated in the following figure.

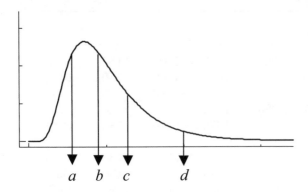

Fig. 3—Any distribution is partitioned into n+1 compartments with a sample of n.

In Figure 3, the sample size is $n=4$; the sample values are denoted as a, b, c and d. The sample presumably comes from some distribution as the figure suggests. Although we do not know the exact locations that the sample values correspond to along the true distribution, we observe that the four values divide the distribution into 5 roughly equal compartments. Each compartment will contain some percentage of the area under the curve so that the sum of each of the percentages is 100%. Assuming that each compartment contains the same area, the probability a value will fall into any compartment is $100[1/(n+1)]\%$.

Similarly, we can compute the percentile that each value represents by $100[i/(n+1)]\%$, where $i = 1, 2, ..., n$. If we ask what percentile is the first order statistic among the four values, we estimate the answer as the $100[1/(4+1)]\% = 20\%$, or 20th percentile. This is because, on average, each of the compartments in Figure 3 will include approximately 20% of the distribution. Since there are $n+1=4+1=5$ compartments in the figure, each compartment is worth 20%. The generalization is obvious. For a sample of n values, the percentile corresponding to the ith order statistic is $100[i/(n+1)]\%$, where $i = 1, 2, ..., n$.

For example, if $n=24$ and we want to know which percentiles are best represented by the 1st and 24th order statistics, we can calculate the percentile for each order statistic. For $X_{(1)}$, the percentile is $100(1)/(24+1) = 4$th; and for $X_{(24)}$, the percentile is $100(24/(24+1) = 96$th. For the illustration in Figure 3, the point a corresponds to the 20th percentile, point b to the 40th percentile, point c to the 60th percentile and point d to the 80th percentile. It is not difficult to extend this application. From the figure it appears that the interval defined by $a \leq x \leq d$ should enclose, on average, 60% of the distribution of X.

We now extend these ideas to estimate the distribution percentiles. For the coating weights in Table 2(b), the sample size is $n=100$. The estimate of the 50th percentile, or sample median, is the number lying halfway between the 50th and 51st order statistics ($X_{(50)} = 1.537$ and $X_{(51)} = 1.543$, respectively). Thus, the sample median is $(1.537 + 1.543)/2 = 1.540$. Note that the middlemost values may be the same (tie). When the sample size is an even number, the sample median will always be taken as halfway between the middle two order statistics. Thus, if the sample size is 250, the median is taken as $(X_{(125)}+X_{(126)})/2$. If the sample size is an odd number, the median is taken as the middlemost order statistic. For example, if the sample size is 13, the sample median is taken as $X_{(7)}$. Note that for an odd numbered sample size, n, the index corresponding to the median will be $i = (n+1)/2$.

We can generalize the estimation of any percentile by using the following convention. Let p be a proportion, so that for the 50th percentile p equals 0.50, for the 25th percentile $p = 0.25$, for the 10th percentile $p = 0.10$, and so forth. To specify a percentile we need only specify p. An estimated percentile will correspond to an order statistic or weighted average of two adjacent order statistics. First, compute an approximate rank using the formula $i = (n+1)p$. If i is an integer then the $100p^{th}$ percentile is estimated as $X_{(i)}$ and we are done. If i is not an integer, then drop the decimal portion and keep the integer portion of i. Let k be the retained integer portion and r be the dropped decimal portion (note: $0<r<1$).

TABLE 2 Continued.

(b) Weight of Coating, oz/ft^2 (data of Table 1 (b))					(c) Breaking Strength, lb. (data of Table 1 (c))
1.323	1.457	1.513	1.567	1.620	568
1.323	1.457	1.513	1.567	1.623	570
1.337	1.460	1.520	1.570	1.627	570
1337	1.467	1.530	1.573	1.633	570
1.337	1.467	1.530	1.573	1.637	572
1.350	1.470	1.533	1.577	1.637	572
1.373	1.473	1.533	1.577	1.637	572
1.373	1.473	1.533	1.577	1.647	576
1.377	1.473	1.537	1.580	1.647	578
1.383	1.477	1.537	1.593	1.647	584
1.383	1.477	1.543	1.600	1.660	
1.393	1.477	1.543	1.600	1.670	
1.420	1.480	1.550	1.600	1.690	
1.420	1.483	1.550	1.603	1.700	
1.423	1.490	1.550	1.603	1.717	
1.433	1.493	1.550	1.603	1.730	
1.437	1.497	1.557	1.603	1.750	
1.440	1.500	1.563	1.607	1.753	
1.443	1.503	1.563	1.617	1.763	
1.450	1.503	1.563	1.620	1.767	

The estimated $100p^{th}$ percentile is computed from the formula $X_{(k)} + r(X_{(k+1)} - X_{(k)})$.

Consider the transverse strengths with $n=270$ and let us find the 2.5th and 97.5th percentiles. For the 2.5th percentile, $p = 0.025$. The approximate rank is computed as $i = (270+1)\ 0.025 = 6.775$. Since this is not an integer, we see that $k=6$ and $r=0.775$. Thus, the 2.5th percentile is estimated by $X_{(6)} + r(X_{(7)}-X_{(6)})$, which is $650 + 0.775(660-650) = 657.75$. For the 97.5th percentile, the approximate rank is $i = (270+1)\ 0.975 = 264.225$. Here again, i is not an integer and so we use $k=264$ and $r=0.225$; however; notice that both $X_{(264)}$ and $X_{(265)}$ are equal to 1400. In this case, the value 1400 becomes the estimate.

GROUPED FREQUENCY DISTRIBUTIONS

7. Introduction

Merely grouping the data values may condense the information contained in a set of observations. Such grouping involves some loss of information but is often useful in presenting engineering data. In the following sections, both tabular and graphical presentation of grouped data will be discussed.

8. Definitions

A grouped frequency distribution of a set of observations is an arrangement that shows the frequency of occurrence of the values of the variable in ordered classes.

The interval, along the scale of measurement, of each ordered class is termed a *bin*.

The *frequency* for any bin is the number of observations in that bin. The frequency for a bin divided by the total number of observations is the *relative frequency* for that bin.

Table 3 illustrates how the three sets of observations given in Table 1 may be organized into grouped frequency distributions. The recommended form of presenting tabular distributions is somewhat more compact, however, as shown in Table 4. Graphical presentation is used in Fig. 4 and discussed in detail in Section 14.

9. Choice of Bin Boundaries

It is usually advantageous to make the bin intervals equal. It is recommended that, in general, the *bin boundaries* be chosen half-way between two possible observations. By choosing bin boundaries in this way, certain difficulties of classification and computation are avoided (See Ref. **2**, pp. 73-76). With this choice, the bin boundary values will usually have one more significant figure (usually a 5) than the values in the original data. For example, in Table 3(*a*), observations were recorded to the nearest 10 psi, hence the bin boundaries were placed at 225, 375, etc., rather than at 220, 370, etc., or 230, 380, etc. Likewise, in Table 3(*b*), observations were recorded to the nearest 0.01 oz/ft^2, hence bin boundaries were placed at 1.275, 1.325, etc., rather than at 1.28, 1.33, etc.

10. Number of Bins

The number of bins in a frequency distribution should preferably be between 13 and 20. (For a discussion of this point, See Ref. **1**, p. 69, and Ref. **18**, pp. 9-12.) Sturge's rule is to make the number of bins equal to $1+3.3\log_{10}(n)$. If the number of observations is, say, less than 250, as few as 10 bins may be of use. When the number of observations is less than 25, a frequency distribution of the data is generally of little value from a presentation standpoint, as for example the 10 observations in Table 3(c). In general, the outline of a frequency distribution when presented graphically is more irregular when the number of bins is larger. This tendency is illustrated in Fig. 4.

11. Rules for Constructing Bins

After getting the ungrouped whole number distribution, one can use a number of popular computer programs to automatically construct a histogram. For example, a spreadsheet program, e.g., Excel[1], can be used by selecting the Histogram item from the Analysis Toolpack menu. Alternatively, you can do it manually by applying the following rules:

- The number of bins (or "cells" or "levels") is set equal to NL = CEIL(2.1 log(n)), where n is the sample size and CEIL is an Excel spreadsheet function that extracts the largest integer part of a decimal number, e.g., 5 is CEIL(4.1)).

- Compute the bin interval as LI = CEIL(RG/NL), where RG = LW-SW, and LW is the largest whole number and SW is the smallest among the *n* observations.

- Find the stretch adjustment as SA = CEIL((NL*LI-RG)/2). Set the start boundary at START = SW-SA-0.5 and then add LI successively NL times to get the bin boundaries. Average successive pairs of boundaries to get the bin midpoints.

---

[1] Excel is a trademark of Microsoft Corporation.

TABLE 3. Three examples of grouped frequency distribution, showing bin midpoints and bin boundaries.

	Bin Midpoint	Bin Boundaries	Observed Frequency
(a) Transverse strength, psi	----------------	225	----------------
(data of Table 1 (*a*))	300	----------------	1
	----------------	375	----------------
	450	----------------	1
	----------------	525	----------------
	600	----------------	6
	----------------	675	----------------
	750	----------------	38
	----------------	825	----------------
	900	----------------	80
	----------------	975	----------------
	1050	----------------	83
	----------------	1125	----------------
	1200	----------------	39
	----------------	1275	----------------
	1350	----------------	17
	----------------	1425	----------------
	1500	----------------	2
	----------------	1575	----------------
	1650	----------------	2
	----------------	1725	----------------
	1800	----------------	0
	----------------	1875	----------------
	1950	----------------	1
	----------------	2025	----------------
	Total		270
(b) Weight of coating, oz/ft^2	----------------	1.275	----------------
(data of Table 1 (*b*))	1.300	----------------	2
	----------------	1.325	----------------
	1.350	----------------	6
	----------------	1.375	----------------
	1.400	----------------	7
	----------------	1.425	----------------
	1.450	----------------	14
	----------------	1.475	----------------
	1.500	----------------	14
	----------------	1.525	----------------
	1.550	----------------	22
	----------------	1.575	----------------
	1.600	----------------	17
	----------------	1.625	----------------
	1.650	----------------	10
	----------------	1.675	----------------
	1.700	----------------	3
	----------------	1.725	----------------
	1.750	----------------	5
	----------------	1.775	----------------
	Total		100
(c) Breaking strength, lb	----------------	567	----------------
(data Table 1 (*c*))	568	----------------	1
	----------------	569	----------------
	570	----------------	3
	----------------	571	----------------
	572	----------------	3
	----------------	573	----------------
	574	----------------	0
	----------------	575	----------------
	576	----------------	1
	----------------	577	----------------
	578	----------------	1
	----------------	579	----------------
	580	----------------	0
	----------------	581	----------------
	582	----------------	0
	----------------	583	----------------
	584	----------------	1
	----------------	585	----------------
	Total		10

TABLE 4. Four methods of presenting a tabular frequency distribution (data of **TABLE 1(a)**).

(a) Frequency		(b) Relative Frequency (expressed in percentages)	
Transverse Strength, psi	Number of Bricks Having Strength Within Given Limits	Transverse Strength, psi	Percentage of Bricks Having Strength Within Given Limits
225 to 375	1	225 to 375	0.4
375 to 525	1	375 to 525	0.4
525 to 675	6	525 to 675	2.2
675 to 825	38	675 to 825	14.1
825 to 975	80	825 to 975	29.6
975 to 1125	83	975 to 1125	30.7
1125 to 1275	39	1125 to 1275	14.5
1275 to 1425	17	1275 to 1425	6.3
1425 to 1575	2	1425 to 1575	0.7
1575 to 1725	2	1575 to 1725	0.7
1725 to 1875	0	1725 to 1875	0.0
1875 to 2025	1	1875 to 2025	0.4
Total	270	Total	100.0
		Number of observations = 270	

(c) Cumulative Frequency		(d) Cumulative Relative Frequency (expressed in percentages)	
Transverse Strength, psi	Number of Bricks Having Strength less than Given Values	Transverse Strength, psi	Percentage of Bricks Having Strength less than Given Values
375	1	375	0.4
525	2	525	0.8
675	8	675	3.0
825	46	825	17.1
975	126	975	46.7
1125	209	1125	77.4
1275	248	1275	91.9
1425	265	1425	98.2
1575	267	1575	98.9
1725	269	1725	99.6
1875	269	1875	99.6
2025	270	2025	100.0
		Number of observations = 270	

NOTE—"Number of observations" should be recorded with tables of relative frequencies.

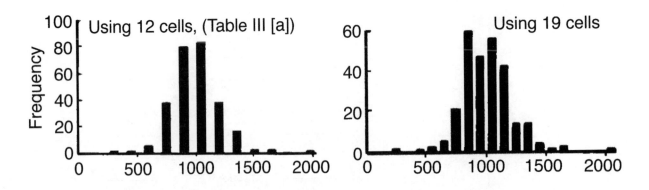

Fig. 4—Illustrating increased irregularity with larger number of cells, or bins.

- Having defined the bins, the last step is to count the whole numbers in each bin and thus record the grouped frequency distribution as the bin midpoints with the frequencies in each.

- The user may improve upon the rules but they will produce a useful starting point and do obey the general principles of construction of a frequency distribution.

Figure 5 illustrates a convenient method of classifying observations into bins when the number of observations is not large. For each observation, a mark is entered in the proper bin. These marks are grouped in five's as the tallying proceeds, and the completed tabulation itself, if neatly done, provides a good picture of the frequency distribution.

If the number of observations is, say, over 250, and accuracy is essential, the use of a computer may be preferred.

12. Tabular Presentation

Methods of presenting tabular frequency distributions are shown in Table 4. To make a frequency tabulation more understandable, relative frequencies may be listed as well as actual frequencies. If only relative frequencies are given, the table cannot be regarded as complete unless the total number of observations is recorded.

Confusion often arises from failure to record bin boundaries correctly. Of the four methods, A to D, illustrated for strength measurements made *to the nearest 10 lb.*, only Methods A and B are recommended (Table 5). Method C gives no clue as to how observed values of 2100, 2200, etc., which fell exactly at bin boundaries were classified. If such values were consistently placed in the next higher bin, the real bin boundaries are those of Method A. Method D is liable to misinterpretation since strengths were measured to the nearest 10 lb. only.

Transverse Strength, psi.		Frequency
225 to 375	|	1
375 to 525	|	1
525 to 675	ЖІ	6
675 to 825	Ж Ж Ж Ж Ж Ж Ж III	38
825 to 975	ЖЖЖЖЖЖЖЖЖЖЖЖЖЖЖЖ	80
975 to 1125	ЖЖЖЖЖЖЖЖЖЖЖЖЖЖЖЖЖ III	83
1125 to 1275	ЖЖЖЖЖЖЖ IIII	39
1275 to 1425	ЖЖЖ II	17
1425 to 1575	II	2
1575 to 1775	II	2
1725 to 1875		0
1875 to 2025	|	1
	Total	270

Fig. 5—Method of classifying observations. Data of Table 1(a).

TABLE 5. Methods A through D illustrated for strength measurements to the nearest 10 lb.

	RECOMMENDED				NOT RECOMMENDED			
	METHOD A		METHOD B		METHOD C		METHOD D	
STRENGTH, lb.	NUMBER OF OBSER-VATIONS	STRENGTH, lb.	NUMBER OF OBSER-VATIONS	STRENGTH, lb.	NUMBER OF OBSER-VATIONS	STRENGTH, lb.	NUMBER OF OBSER-VATIONS
1995 to 2095	1	2000 to 2090	1	2000 to 2100	1	2000 to 2099	1
2095 to 2195	3	2100 to 2190	3	2100 to 2200	3	2100 to 2199	3
2195 to 2295	17	2200 to 2290	17	2200 to 2300	17	2200 to 2299	17
2295 to 2395	36	2300 to 2390	36	2300 to 2400	36	2300 to 2399	36
2395 to 2495	82	2400 to 2490	82	2400 to 2500	82	2400 to 2499	82
etc.	etc.	etc.	etc.	etc.	etc.	etc.	etc.

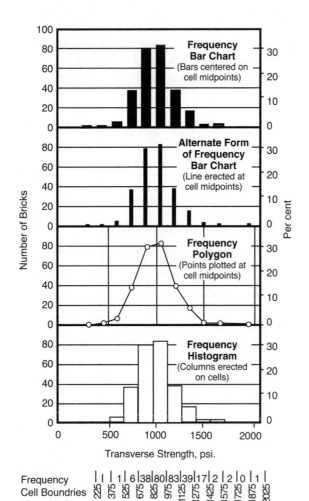

FIG. 6—Graphical presentations of a frequency distribution. Data of Table 1(a) as grouped in Table 3(a).

13. Graphical Presentation

Using a convenient horizontal scale for values of the variable and a vertical scale for bin frequencies, frequency distributions may be reproduced graphically in several ways as shown in Fig. 6. The *frequency bar chart* is obtained by erecting a series of bars, centered on the bin midpoints, with each bar having a height equal to the bin frequency. An alternate form of frequency bar chart may be constructed by using lines rather than bars. The distribution may also be shown by a series of points or circles representing bin frequencies plotted at bin midpoints. The *frequency polygon* is obtained by joining these points by straight lines. Each endpoint is joined to the base at the next bin midpoint to close the polygon.

Another form of graphical representation of a frequency distribution is obtained by placing along the graduated horizontal scale a series of vertical columns, each having a width equal to the bin width and a height equal to the bin frequency. Such a graph, shown at the bottom of Fig. 6, is called the *frequency histogram* of the distribution. In the histogram, if bin widths are arbitrarily given the value 1, the area enclosed by the steps represents frequency exactly, and the sides of the columns designate bin boundaries.

The same charts can be used to show relative frequencies by substituting a relative frequency scale, such as that shown in Fig. 6. It is often advantageous to show both a frequency scale and a relative frequency scale. If only a relative frequency scale is given on a chart, the number of observations should be recorded as well.

14. Cumulative Frequency Distribution

Two methods of constructing cumulative frequency polygons are shown in Fig. 7. Points are plotted at bin boundaries. The upper chart gives cumulative frequency and relative cumulative frequency plotted on an arithmetic scale. This type of graph is often called an *ogive* or "s" graph. Its use is discouraged mainly because it is usually difficult to interpret the tail regions.

The lower chart shows a preferable method by plotting the relative cumulative frequencies on a normal probability scale. A Normal distribution (see Fig. 14) will plot cumulatively as a straight line on this scale. Such graphs can be drawn to show the number of observations either "less than" or "greater than" the scale values. (Graph paper with one dimension graduated in terms of the summation of Normal law distribution has been described in Refs. **3,18**). It should be noted that the cumulative percents need to be adjusted to avoid cumulative percents from equaling or exceeding 100%. The probability scale only reaches to 99.9% on most available probability plotting papers. Two methods which will work for estimating cumulative percentiles are [cumulative frequency/(n+1)], and [(cumulative frequency – 0.5)/n].

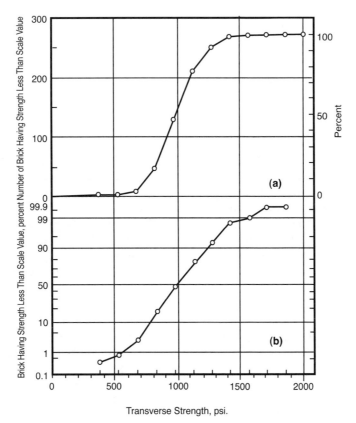

(a) Using arithmetic scale for frequency.
(b) Using probability scale for relative frequency.

Fig. 7—Graphical presentations of a cumulative frequency distribution. Data of Table 4: (a) using arithmetic scale for frequency, and (b) using probability scale for relative frequency.

For some purposes, the number of observations having a value "less than" or "greater than" particular scale values is of more importance than the frequencies for particular bins. A table of such frequencies is termed a *cumulative frequency distribution*. The "less than" cumulative frequency distribution is formed by recording the frequency of the first bin, then the sum of the first and second bin frequencies, then the sum of the first, second, and third bin frequencies, and so on.

Because of the tendency for the grouped distribution to become irregular when the number of bins increases, it is sometimes preferable to calculate percentiles from the cumulative frequency distribution rather than from the order statistics. This is recommended as *n* passes the hundreds and reaches the thousands of observations. The method of calculation can easily be illustrated geometrically by using Table 4(d), Cumulative Relative Frequency and the problem of getting the 2.5th and 97.5th percentiles.

We first define the *cumulative relative frequency function*, *F(x)*, from the bin boundaries and the cumulative relative frequencies. It is just a sequence of straight lines connecting the points (X=235, F(235)=0.000), (X=385, F(385)=0.0037), (X=535, F(535)=0.0074), and so on up to (X=2035, F(2035)=1.000). Notice in Fig. 7, with an arithmetic scale for percent, and you can see the function. A horizontal line at height 0.025 will cut the curve between X=535 and X=685, where the curve rises from 0.0074 to 0.0296. The full vertical distance is 0.0296-0.0074 = 0.0222, and the portion lacking is 0.0250-0.0074 = 0.0176, so this cut will occur at (0.0176/0.0222) 150+535 = 653.9 psi. The horizontal at 97.5% cuts the curve at 1419.5 psi.

15. "Stem and Leaf" Diagram

It is sometimes quick and convenient to construct a "stem and leaf" diagram, which has the appearance of a histogram turned on its side. This kind of diagram does not require choosing explicit bin widths or boundaries.

The first step is to reduce the data to two or three-digit numbers by: (1) dropping constant initial or final digits, like the final zero's in Table 1(a) or the initial one's in Table 1(b); (2) removing the decimal points; and finally, (3) rounding the results after (1) and (2), to two or three-digit numbers we can call coded observations. For instance, if the initial one's and the decimal points in the data of Table 1(b) are dropped, the coded observations run from 323 to 767, spanning 445 successive integers.

If forty successive integers per class interval are chosen for the coded observations in this example, there would be 12 intervals; if thirty successive integers, then 15 intervals; and if twenty successive integers then 23 intervals. The choice of 12 or 23 intervals is outside of the recommended interval from 13 to 20. While either of these might nevertheless be chosen for convenience, the flexibility of the stem and leaf procedure is best shown by choosing thirty successive integers per interval, perhaps the least convenient choice of the three possibilities.

Each of the resulting 15 class intervals for the coded observations is distinguished by a first digit and a second. The third digits of the coded observations do not indicate to which intervals they belong and are therefore not needed to construct a stem and leaf diagram in this case. But the first digit may change (by one) within a single class interval. For instance, the first class interval with coded observations beginning with 32, 33 or 34 may be identified by 3(234) and the second class interval by 3(567), but the third class interval includes coded observations with leading digits 38, 39 and 40. This interval may be identified by 3(89)4(0) The intervals, identified in this manner, are listed in the left column of Fig. 8. Each coded observation is set down in turn to the right of its class interval identifier in the diagram using as a symbol its second digit, in the order (from left to right) in which the original observations occur in Table 1(b).

In spite of the complication of changing some first digits within some class intervals, this stem and leaf diagram is quite simple to construct. In this particular case, the diagram reveals "wings" at both ends of the diagram.

First (and second) Digit:	Second Digits Only
3(234)	3 2 2 3 3
3(567)	7 7 7 5
3(89)4(0)	8 9 8
4(123)	2 2 3 3 2
4(456)	6 6 5 5 4 5 4 6
4(789)	7 9 8 7 8 7 7 9 7 9 7 7
5(012)	2 1 0 1 0 0
5(345)	5 3 3 3 3 4 5 5 5 3 4 3 3 5
5(678)	6 7 7 7 7 6 8 6 6 7 7 6
5(9)6(01)	0 0 0 0 9 0 0 1 0
6(234)	2 3 2 4 2 3 4 2 3 3 4
6(567)	6 7
6(89)7(0)	0 9
7(123)	3 1
7(456)	6 5 6 5

FIG. 8—Stem and leaf diagram of data from Table 1(b) with groups based on triplets of first and second decimal digits.

As this example shows, the procedure does not require choosing a precise class interval width or boundary values. At least as important is the protection against plotting and counting errors afforded by using clear, simple numbers in the construction of the diagram—a histogram on its side. For further information on stem and leaf diagrams see Refs. **4** and **18**.

16. "Ordered Stem and Leaf" Diagram and Box Plot

The stem and leaf diagram can be extended to one that is *ordered*. The ordering pertains to the ascending sequence of values within each "leaf". The purpose of ordering the leaves is to make the determination of the *quartiles* an easier task. The quartiles represent the 25th, 50th (median), and 75th percentiles of the frequency distribution. They are found by the method discussed in Section 6.

In Fig. 8a, the quartiles for the data are **bold** and underlined. The quartiles are used to construct another graphic called a *box plot*.

The "box" is formed by the 25th and 75th percentiles, the center of the data is dictated by the 50th percentile (median) and "whiskers" are formed by extending a line from either side of the box to the minimum, $X_{(1)}$ point, and to the maximum, $X_{(n)}$ point. Fig. 8b shows the box plot for the data from Table 1(b). For further information on boxplots, see Ref. **18**.

First (and second) Digit:	Second Digits Only
3(234)	2 2 3 3 3
3(567)	5 7 7 7
3(89)4(0)	8 8 9
4(123)	2 2 2 3 3
4(456)	4 4 5 5 5 6 6 **6**
4(789)	7 7 7 7 7 7 8 8 9 9 9
5(012)	0 0 0 1 1 2
5(345)	3 3 3 3 3 3 **3** 4 4 5 5 5 5 5
5(678)	6 6 6 6 6 7 7 7 7 7 7 8
5(9)6(01)	9 0 0 **0** 0 0 0 0 1
6(234)	2 2 2 2 3 3 3 3 4 4 4
6(567)	6 7
6(89)7(0)	9 0
7(123)	1 3
7(456)	5 5 6 6

FIG. 8a—Ordered stem and leaf diagram of data from Table 1(b) with groups based on triplets of first and second decimal digits. The 25th, 50th and 75th quartiles are shown in bold type and are underlined.

1.323   1.767
     1.4678   1.540   1.6030

FIG. 8b—Box plot of data from Table 1(b)

The information contained in the data may also be summarized by presenting a tabular grouped frequency distribution, if the number of observations is large. A graphical presentation of a distribution makes it possible to visualize the nature and extent of the observed variation.

While some condensation is effected by presenting grouped frequency distributions, further reduction is necessary for most of the uses that are made of ASTM data. This need can be fulfilled by means of a few simple functions of the observed distribution, notably, the *average* and the *standard deviation*.

FUNCTIONS OF A FREQUENCY DISTRIBUTION

17. Introduction

In the problem of condensing and summarizing the information contained in the frequency distribution of a sample of observations, certain functions of the distribution are useful. For some purposes, a statement of the relative frequency within stated limits is all that is needed. For most purposes, however, two salient characteristics of the distribution which are illustrated in Fig. 9a are: (*a*) the position on the scale of measurement—the value about which the observations have a tendency to center, and

(*b*) the spread or dispersion of the observations about the central value.

A third characteristic of some interest, but of less importance, is the skewness or lack of symmetry—the extent to which the observations group themselves more on one side of the central value than on the other (see Fig. 9b).

A fourth characteristic is "kurtosis" which relates to the tendency for a distribution to have a sharp peak in the middle and excessive frequencies on the tails as compared with the Normal distribution or conversely to be relatively flat in the middle with little or no tails (see Fig. 10).

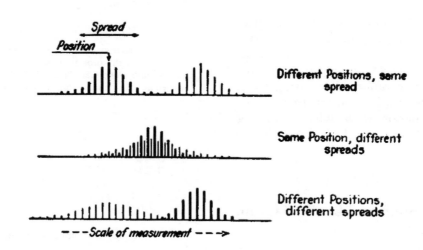

FIG. 9a—Illustrating two salient characteristics of distributions—position and spread.

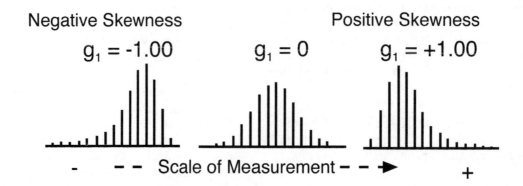

FIG. 9b—Illustrating a third characteristic of frequency distributions—skewness, and particular values of skewness, g_1.

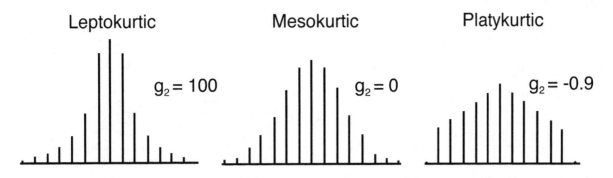

FIG. 10—Illustrating the kurtosis of a frequency distribution and particular values of g_2

Several representative sample measures are available for describing these characteristics, but by far the most useful are the arithmetic mean $\overline{X}$, the standard deviation s, the skewness factor g_1, and the kurtosis factor g_2—all algebraic functions of the observed values. Once the numerical values of these particular measures have been determined, the original data may usually be dispensed with and two or more of these values presented instead.

The four characteristics of the distribution of a sample of observations just discussed are most useful when the observations form a single heap with a single peak frequency not located at either extreme of the sample values. If there is more than one peak, a tabular or graphical representation of the frequency distribution conveys information the above four characteristics do not.

18. Relative Frequency

The relative frequency p within stated limits on the scale of measurement is the ratio of the number of observations lying within those limits to the total number of observations.

In practical work, this function has its greatest usefulness as a measure of *fraction nonconforming*, in which case it is the fraction, p, representing the ratio of the number of observations lying outside specified limits (or beyond a specified limit) to the total number of observations.

19. Average (Arithmetic Mean)

The average (arithmetic mean) is the most widely used measure of central tendency. The term *average* and the symbol $\overline{X}$ will be used in this Manual to represent the arithmetic mean of a sample of numbers.

The average, $\overline{X}$, of a sample of n numbers, $X_1, X_2, ..., X_n$, is the sum of the numbers divided by n, that is

$$\overline{X} = \frac{X_1 + X_2 + \cdots + X_n}{n} = \frac{\sum_{i=1}^{n} X_i}{n} \qquad (1)$$

where the expression $\sum_{i=1}^{n} X_i$ means "the sum of all values of X, from X_1 to X_n, inclusive."

Considering the n values of X as specifying the positions on a straight line of n particles of equal weight, the average corresponds to the center of gravity of the system. The average of a series of observations is expressed in the same units of measurement as the observations, that is, if the observations are in pounds, the average is in pounds.

20. Other Measures of Central Tendency

The *geometric mean*, of a sample of n numbers, $X_1, X_2,..., X_n$, is the nth root of their product, that is

$$\text{geometric mean} = \sqrt[n]{X_1 X_2 \cdots X_n} \qquad (2)$$

or

$$\log (\text{geometric mean})$$

$$= \frac{\log X_1 + \log X_2 + \cdots + \log X_n}{n} \qquad (3)$$

Equation 3, obtained by taking logarithms of both sides of Eq 2, provides a convenient method for computing the geometric mean using the logarithms of the numbers.

NOTE

The distribution of some quality characteristics is such that a transformation, using logarithms of the observed values, gives a substantially Normal distribution. When this is true, the transformation is distinctly advantageous for (in accordance with Section 29) much of the total information can be presented by two functions, the average, $\overline{X}$, and the standard deviation, s, of the logarithms of the observed values. The problem of transformation is, however, a complex one that is beyond the scope of this Manual. See Ref. **18**.

The *median* of the frequency distribution of n numbers is the middlemost value.

The *mode* of the frequency distribution of n numbers is the value that occurs most frequently. With grouped data, the mode may vary due to the choice of the interval size and the starting points of the bins.

21. Standard Deviation

The standard deviation is the most widely used measure of dispersion for the problems considered in **PART 1** of the Manual.

For a sample of n numbers, X_1, X_2..., X_n, the sample standard deviation is commonly defined by the formula

$$s = \sqrt{\frac{(X_1 - \overline{X})^2 + (X_2 - \overline{X})^2 + \cdots + (X_n - \overline{X})^2}{n-1}}$$

$$= \sqrt{\frac{\sum_{i=1}^{n}(X_i - \overline{X})^2}{n-1}} \qquad (4)$$

where $\overline{X}$ is defined by Eq 1. The quantity s^2 is called the sample *variance*.

The standard deviation of any series of observations is expressed in the same units of measurement as the observations, that is, if the observations are in pounds, the standard deviation is in pounds. (Variances would be measured in pounds squared.)

A frequently more convenient formula for the computation of s is

$$s = \sqrt{\frac{\sum_{i=1}^{n} X_i^2 - \frac{\left(\sum_{i=1}^{n} X_i\right)^2}{n}}{n-1}} \qquad (5)$$

but care must be taken to avoid excessive rounding error when n is larger than s.

NOTE

A useful quantity related to the standard deviation is the *root-mean-square deviation*

$$s_{(rms)} = \sqrt{\frac{\sum_{i=1}^{n}\left(X - \overline{X}\right)^2}{n}} = s\sqrt{\frac{n-1}{n}}$$

22. Other Measures of Dispersion

The *coefficient of variation, cv*, of a sample of n numbers, is the ratio (sometimes the coefficient is expressed as a percentage) of their standard deviation, s, to their average $\overline{X}$. It is given by

$$cv = \frac{s}{\overline{X}} \qquad (6)$$

The coefficient of variation is an adaptation of the standard deviation, which was developed by Prof. Karl Pearson to express the variability of a set of numbers on a relative scale rather than on an absolute scale. It is thus a dimensionless number. Sometimes it is called the *relative standard deviation*, or *relative error*.

The *average deviation* of a sample of n numbers, X_1, X_2, ..., X_n, is the average of the absolute values of the deviations of the numbers from their average $\overline{X}$ that is

$$\text{average deviation} = \frac{\sum_{i=1}^{n} |X_i - \overline{X}|}{n} \qquad (7)$$

where the symbol $|\ |$ denotes the absolute value of the quantity enclosed.

The *range R* of a sample of n numbers is the difference between the largest number and the smallest number of the sample. One computes R from the order statistics as R = $X_{(n)}$-$X_{(1)}$. This is the simplest measure of dispersion of a sample of observations.

23. Skewness—g_1

A useful measure of the lopsidedness of a sample frequency distribution is the coefficient of skewness g_1.

The coefficient of skewness g_1, of a sample of n numbers, X_1, X_2, ..., X_n, is defined by the expression $g_1 = k_3/s^3$. Where k_3 is the third k-statistic as defined by R. A. Fisher. The k-statistics were devised to serve as the moments of small sample data. The first moment is the mean, the second is the variance, and the third is the average of the cubed deviations and so on. Thus, $k_1 = \overline{X}$, $k_2 = s^2$,

$$k_3 = \frac{n \sum (X_i - \overline{X})^3}{(n-1)(n-2)}$$

Notice that when n is large

$$g_1 = \frac{\sum_{i=1}^{n} (X_i - \overline{X})^3}{ns^3} \qquad (8)$$

This measure of skewness is a pure number and may be either positive or negative. For a symmetrical distribution, g_1 is zero. In general, for a nonsymmetrical distribution, g_1 is negative if the long tail of the distribution extends to the left, towards smaller values on the scale of measurement, and is positive if the long tail extends to the right, towards larger values on the scale of measurement. Figure 9 shows three unimodal distributions with different values of g_1.

23a. Kurtosis—g_2

The peakedness and tail excess of a sample frequency distribution is generally measured by the coefficient of kurtosis g_2.

The coefficient of kurtosis g_2 for a sample of n numbers, X_1, X_2, ..., X_n, is defined by the expression $g_2 = k_4/s^4$ and

$$k_4 = \frac{n(n+1)\sum (X_i - \overline{X})^4}{(n-1)(n-2)(n-3)} - \frac{3(n-1)^2 s^4}{(n-2)(n-3)}$$

Notice that when n is large

$$g_2 = \frac{\sum_{i=1}^{n} (X_1 - \overline{X})^4}{ns^4} - 3 \qquad (9)$$

Again this is a dimensionless number and may be either positive or negative. Generally, when a distribution has a sharp peak, thin shoulders, and small tails relative to the bell-shaped distribution characterized by the Normal distribution, g_2 is positive. When a distribution is flat-topped with fat tails, relative to the Normal distribution, g_2 is negative. Inverse relationships do not necessarily follow. We cannot definitely infer anything about the shape of a distribution from knowledge of g_2 unless we are willing to assume some theoretical curve, say a Pearson curve, as being appropriate as a graduation formula (see Fig. 14 and Section 30). A distribution with a positive g_2 is said to be *leptokurtic*. One with a negative g_2 is said to be *platykurtic*. A distribution with $g_2 = 0$ is said to be *mesokurtic*. Figure 10 gives three unimodal distributions with different values of g_2.

24. Computational Tutorial

The method of computation can best be illustrated with an artificial example for n=4 with $X_1 = 0$, $X_2 = 4$, $X_3 = 0$, and $X_4 = 0$. Please first verify that $\overline{X} = 1$. The deviations from this mean are found as –1, 3, -1, and –1. The sum of the squared deviations is thus 12 and $s^2 = 4$. The sum of cubed deviations is $-1+27-1-1 = 24$, and thus $k_3 = 16$. Now we find $g_1 = 16/8 = 2$. Please verify that $g_2 = 4$. Since both g_1 and g_2 are positive, we can say that the distribution is both skewed to the right and leptokurtic relative to the Normal distribution.

Of the many measures that are available for describing the salient characteristics of a sample frequency distribution, the average $\overline{X}$, the standard deviation s, the skewness g_1, and the kurtosis g_2, are particularly useful for summarizing the information contained therein. So long as one uses them only as rough indications of uncertainty we list approximate sampling standard deviations of the quantities $\overline{X}$, s^2, g_1 and g_2, as

$$SE\left(\overline{X}\right) = s/\sqrt{n} \,,$$

$$SE\left(s\right) = s^2/\sqrt{4n} \,,$$

$$SE\left(g_1\right) = \sqrt{6/n} \,, \text{ and}$$

$$SE\left(g_2\right) = \sqrt{24/n} \,, \text{ respectively.}$$

When using a computer software calculation, the ungrouped whole number distribution values will lead to less round off in the printed output and are simple to scale back to original units. The results for the data from Table 2 are given in Table 6.

AMOUNT OF INFORMATION CONTAINED IN p, $\overline{X}$, s, g_1, AND g_2

25. Summarizing the Information

Given a sample of n observations, $X_1, X_2, X_3, ...,$ X_n, of some quality characteristic, how can we present concisely information by means of which the observed distribution can be closely approximated, that is, so that the percentage of the total number, n, of observations lying within any stated interval from, say, $X = a$ to $X = b$, can be approximated?

Table 6. Summary Statistics for Three Sets of Data

Datasets	$\overline{X}$	s	g_1	g_2
Transverse Strength, psi	999.8	201.8	0.611	2.567
Weight of Coating, oz/ft^2	1.535	0.1038	0.013	-0.291
Breaking Strength, lb	573.2	4.826	1.419	1.797

The total information can be presented only by giving all of the observed values. It will be shown, however, that much of the total information is contained in a few simple functions—notably the average $\overline{X}$, the standard deviation s, the skewness g_1, and the kurtosis g_2.

26. Several Values of Relative Frequency, p

By presenting, say, 10 to 20 values of relative frequency p, corresponding to stated bin intervals and also the number n of observations, it is possible to give practically all of the total information in the form of a tabular grouped frequency distribution. If the ungrouped distribution has any peculiarities, however, the choice of bins may have an important bearing on the amount of information lost by grouping.

27. Single Percentile of Relative Frequency, p

If we present but a percentile value, Q_p, of relative frequency p, such as the fraction of the total number of observed values falling outside of a specified limit and also the number n of observations, the portion of the total information presented is very small. This follows from the fact that quite dissimilar distributions may have identically the same percentile value as illustrated in Fig. 11.

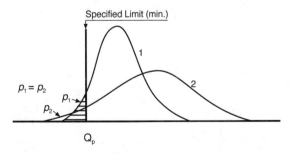

FIG. 11—Quite different distributions may have the same percentile value of _p_, fraction of total observations below specified limit.

NOTE

For the purposes of **PART 1** of this Manual, the curves of Figs. 11 and 12 may be taken to represent frequency histograms with small bin widths and based on large samples. In a frequency histogram, such as that shown at the bottom of Fig. 5, let the percentage relative frequency between any two bin boundaries be represented by the *area* of the histogram between those boundaries, the total area being 100 percent. Since the bins are of uniform width, the relative frequency in any bin is then proportional to the *height* of that bin and may be read on the vertical scale to the right.

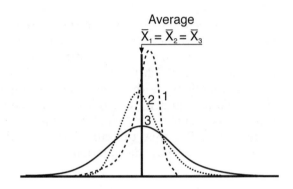

FIG. 12—Quite different distributions may have the same average.

If the sample size is increased and the bin width reduced, a histogram in which the relative frequency is measured by area approaches as a limit the frequency distribution of the population, which in many cases can be represented by a smooth curve. The relative frequency between any two values is then represented by the *area* under the curve and between ordinates erected at those values. Because of the method of generation, the ordinate of the curve may be regarded as a curve of *relative frequency density*. This is analogous to the representation of the variation of density along a rod of uniform cross section by a smooth curve. The weight between any two points along the rod is proportional to the area under the curve between the two ordinates and we may speak of the *density* (that is, weight density) at any point but not of the *weight* at any point.

28. Average $\overline{X}$ Only

If we present merely the average, $\overline{X}$, and number, n, of observations, the portion of the total information presented is very small. Quite dissimilar distributions may have identically the same value of $\overline{X}$ as illustrated in Fig. 12.

In fact, no single one of the five functions, Q_p, $\overline{X}$, s, g_1, or g_2, presented alone, is generally capable of giving much of the total information in the original distribution. Only by presenting two or three of these functions can a fairly complete description of the distribution generally be made.

An exception to the above statement occurs when theory and observation suggest that the underlying law of variation is a distribution for which the basic characteristics are all functions of the mean. For example, "life" data "under controlled conditions" sometimes follows a negative exponential distribution. For this, the cumulative relative frequency is given by the equation

$$F(X) = 1 - e^{-x/\theta} \qquad 0 < X < \infty \qquad (14)$$

This is a single parameter distribution for which the mean and standard deviation both equal θ. That the negative exponential distribution is the underlying law of variation can be checked by noting whether values of $1 - F(X)$ for the sample data tend to plot as a straight line on ordinary semi-logarithmic paper. In such a situation, knowledge of $\overline{X}$ will, by taking $\theta = \overline{X}$ in Eq. 14 and using tables

of the exponential function, yield a fitting formula from which estimates can be made of the percentage of cases lying between any two specified values of X. Presentation of $\overline{X}$ and n is sufficient in such cases provided they are accompanied by a statement that there are reasons to believe that X has a negative exponential distribution.

29. Average $\overline{X}$ and Standard Deviation s

These two functions contain some information even if nothing is known about the form of the observed distribution, and contain much information when certain conditions are satisfied. For example, more than $1 - 1/k^2$ of the total number n of observations lie within the closed interval $\overline{X} \pm ks$ (where k is not less than 1).

This is *Chebyshev's inequality* and is shown graphically in Fig. 13. The inequality holds true of *any* set of finite numbers regardless of how they were obtained. Thus if $\overline{X}$ and s are presented, we may say at once that more than 75 percent of the numbers lie within the interval $\overline{X} \pm 2s$; stated in another way, less than 25 percent of the numbers differ from $\overline{X}$ by more than $2s$. Likewise, more than 88.9 percent lie within the interval $\overline{X} \pm 3s$, etc. Table 7 indicates the conformance with Chebyshev's inequality of the three sets of observations given in Table 1.

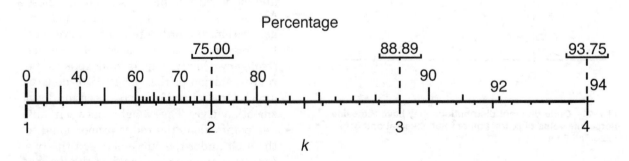

FIG. 13—Percentage of the total observations lying within the interval $\overline{x} \pm ks$ always exceeds the percentage given on this chart.

TABLE 7. Comparison of observed percentages and Chebyshev's minimum percentages of the total observations lying within given intervals.

INTERVAL, $\overline{X} \pm ks$	CHEBYSHEV'S MINIMUM OBSERVATIONS LYING WITHIN THE GIVEN INTERVAL $\overline{X} \pm ks$	OBSERVED PERCENTAGES[a]		
		DATA OF TABLE 1(a) (n = 270)	DATA OF TABLE 1(b) (n = 100)	DATA OF TABLE 1(c) (n = 10)
$\overline{X} \pm 2.0s$	75.0	96.7	94	90
$\overline{X} \pm 2.5s$	84.0	97.8	100	90
$\overline{X} \pm 3.0s$	88.9	98.5	100	100

[a]Data of Table 1(a): $\overline{X} = 1000$, $s = 202$; data of Table 1(b): $\overline{X} = 1.535$, $s = 0.105$; data of Table 1(c): $\overline{X} = 573.2$, $s = 4.58$.

To determine approximately just what percentages of the total number of observations lie within given limits, as contrasted with minimum percentages within those limits, requires additional information of a restrictive nature. If we present $\overline{X}$, s, and n, and are able to add the information "data obtained under controlled conditions," then it is possible to make such estimates satisfactorily for limits spaced equally above and below $\overline{X}$.

What is meant technically by "controlled conditions" is discussed by Shewhart (see Ref. 1) and is beyond the scope of this Manual. Among other things, the concept of control includes the idea of homogeneous data—a set of observations resulting from measurements made under the same essential conditions and representing material produced under the same essential conditions. It is sufficient for present purposes to point out that if data are obtained under "controlled conditions," it may be assumed that the observed frequency distribution can, for most practical purposes, be graduated by some theoretical curve say, by the Normal law or by one of the non-normal curves belonging to the system of frequency curves developed by Karl Pearson. (For an extended discussion of Pearson curves, see Ref. 5). Two of these are illustrated in Fig. 14.

The applicability of the Normal law rests on two converging arguments. One is mathematical and proves that the distribution of a sample mean obeys the Normal law no matter what the shape of the distributions are for each of the separate observations. The other is that experience with many, many sets of data show that more of them approximate the Normal law than any other distribution. In the field of statistics, this effect is known as the *central limit theorem*.

Supposing a smooth curve plus a gradual approach to the horizontal axis at one or both sides derived the Pearson system of curves. The Normal distribution's fit to the set of data may be checked roughly by plotting the cumulative data on Normal probability paper (see Section 13). Sometimes if the original data do not appear to follow the Normal law, some transformation of the data, such as log X, will be approximately normal.

Thus, the phrase "data obtained under controlled conditions" is taken to be the equivalent of the more mathematical assertion that "the functional form of the distribution may be represented by some specific curve." However, conformance of the shape of a frequency distribution with some curve should by no means be taken as a sufficient criterion for control.

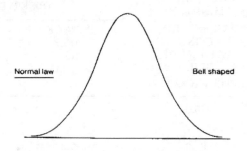

Examples of two Pearson non-normal frequency curves

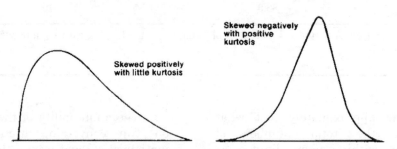

FIG. 14—A frequency distribution of observations obtained under controlled conditions will usually have an outline that conforms to the Normal law or a non-normal Pearson frequency curve.

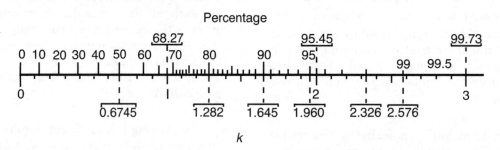

FIG. 15—Normal law integral diagram giving percentage of total area under Normal law curve falling within the range $\mu \pm k\sigma$. This diagram is also useful in probability and sampling problems, expressing the upper (percentage) scale values in decimals to represent "probability."

Generally for controlled conditions, the percentage of the total observations in the original sample lying within the interval $\overline{X} \pm ks$ may be determined approximately from the chart of Fig. 15, which is based on the Normal law integral. The approximation may be expected to be better the larger the number of observations. Table 8 compares the observed percentages of the total number of observations lying within several symmetrical intervals about $\overline{X}$ with those estimated from a knowledge of $\overline{X}$ and s, for the three sets of observations given in Table 1.

30. Average $\overline{X}$, Standard Deviation s, Skewness g_1, and Kurtosis g_2

If the data are obtained under "controlled conditions" and if a Pearson curve is assumed appropriate as a graduation formula, the presentation of g_1 and g_2 in addition to $\overline{X}$ and s will contribute further information. They will give no immediate help in determining the percentage of the total observations lying within a symmetric interval about the average $\overline{X}$, that is, in the interval of $\overline{X} \pm ks$.

TABLE 8. Comparison of observed percentages and theoretical estimated percentages of the total observations lying within given intervals.

INTERVAL, $\overline{X} \pm ks$	THEORETICAL ESTIMATED PERCENTAGES[a] OF TOTAL OBSERVATIONS LYING WITHIN THE GIVEN INTERVAL $\overline{X} \pm ks$	OBSERVED PERCENTAGES		
		DATA OF TABLE 1(a) (n = 270)	DATA OF TABLE 1(b) (n = 100)	DATA OF TABLE 1(c) (n = 10)
$\overline{X} \pm 0.6745s$	50.0	52.2	54	70
$\overline{X} \pm 1.0s$	68.3	76.3	72	80
$\overline{X} \pm 1.5s$	86.6	89.3	84	90
$\overline{X} \pm 2.0s$	95.5	96.7	94	90
$\overline{X} \pm 2.5s$	98.7	97.8	100	90
$\overline{X} \pm 3.0s$	99.7	98.5	100	100

[a]Use Fig. 15 with $\overline{X}$ and s as estimates of μ and σ.

What they do is to help in estimating observed percentages (in a sample already taken) in an interval whose limits are not equally spaced above and below $\overline{X}$.

If a Pearson curve is used as a graduation formula, some of the information given by g_1 and g_2 may be obtained from Table 9 which is taken from Table 42 of the *Biometrika Tables for Statisticians*. For $\beta_1 = g_1^2$ and $\beta_2 = g_2 + 3$, this table gives values of k_L for use in estimating the lower 2.5 percent of the data and values of k_U for use in estimating the upper 2.5 percent point. More specifically, it may be estimated that 2.5 percent of the cases are less than $\overline{X} - k_L s$ and 2.5 percent are greater than $\overline{X} + k_U s$. Put another way, it may be estimated that 95 percent of the cases are between $\overline{X} - k_L s$ and $\overline{X} + k_U s$.

Table 42 of the *Biometrika Tables for Statisticians* also gives values of k_L and k_U for 0.5, 1.0, and 5.0 percent points.

Example

For a sample of 270 observations of the transverse strength of bricks, the sample distribution is shown in Fig. 5. From the sample values of $g_1 = 0.61$ and $g_2 = 2.57$, we take $\beta_1 = g_1^2 = (0.61)^2 = 0.37$ and $\beta_2 = g_2 + 3 = 2.57 + 3 = 5.57$. Thus, from Tables 9(a) and

(b) we may estimate that approximately 95 percent of the 270 cases lie between $\overline{X} - k_L s$ and $\overline{X} + k_U s$, or between $1000 - 1.801 (201.8) = 636.6$ and $1000 + 2.17 (201.8) = 1437.7$. The actual percentage of the 270 cases in this range is 96.3 percent (see Table 2(a)).

Notice that using just $\overline{X} \pm 1.96s$ gives the interval 604.3 to 1395.3 which actually includes 95.9% of the cases versus a theoretical percentage of 95%. The reason we prefer the Pearson curve interval arises from knowing that the $g_1 = 0.63$ value has a standard error of 0.15 $\left(= \sqrt{6/270} \right)$ and is thus about four standard errors above zero. That is, if future data come from the same conditions it is highly probable that they will also be skewed. The 604.3 to 1395.3 interval is symmetric about the mean, while the 636.6 to 1437.7 interval is offset in line with the anticipated skewness. Recall that the interval based on the order statistics was 657.8 to 1400 and that from the cumulative frequency distribution was 653.9 to 1419.5.

When computing the median, all methods will give essentially the same result but we need to choose among the methods when estimating a percentile near the extremes of the distribution.

TABLE 9. Lower and upper 2.5 percent points k_L and k_u of the standardized deviate $(X - \mu)/\sigma$, given by Pearson frequency curves for designated values of β_1 (estimated as equal to g_1^2) and β_2 (estimated as equal to $g_2 + 3$).

	β_1 / β_2	0.00	0.01	0.03	0.05	0.10	0.15	0.20	0.30	0.40	0.50	0.60	0.70	0.80	0.90	1.00
(a) Lower[a] k_L	1.8	1.65	...	...	...	...	...	...	...	...	...	...	...	...	...	...
	2.0	1.76	1.68	1.62	1.56	...	...	...	...	...	...	...	...	...	...	...
	2.2	1.83	1.76	1.71	1.66	1.57	1.49	1.41	...	...	...	...	...	...	...	...
	2.4	1.88	1.82	1.77	1.73	1.65	1.58	1.51	1.39	...	...	...	...	...	...	...
	2.6	1.92	1.86	1.82	1.78	1.71	1.64	1.58	1.47	1.37	...	...	...	...	...	...
	2.8	1.94	1.89	1.85	1.82	1.76	1.70	1.65	1.55	1.45	1.35	...	...	...	...	...
	3.0	1.96	1.91	1.87	1.84	1.79	1.74	1.69	1.60	1.52	1.42	1.33	...	...	...	...
	3.2	1.97	1.93	1.89	1.86	1.81	1.77	1.72	1.65	1.57	1.49	1.40	1.32	1.24	...	...
	3.4	1.98	1.94	1.90	1.88	1.83	1.79	1.75	1.68	1.61	1.54	1.46	1.39	1.31	1.23	...
	3.6	1.99	1.95	1.91	1.89	1.85	1.81	1.77	1.71	1.65	1.58	1.51	1.44	1.38	1.30	1.23
	3.8	1.99	1.95	1.92	1.90	1.86	1.82	1.79	1.73	1.67	1.62	1.56	1.49	1.43	1.36	1.29
	4.0	1.99	1.96	1.93	1.91	1.87	1.84	1.81	1.75	1.70	1.64	1.59	1.53	1.47	1.41	1.35
	4.2	2.00	1.96	1.93	1.91	1.88	1.84	1.82	1.76	1.72	1.67	1.62	1.56	1.51	1.45	1.40
	4.4	2.00	1.96	1.94	1.92	1.88	1.85	1.83	1.78	1.73	1.69	1.64	1.59	1.54	1.49	1.44
	4.6	2.00	1.96	1.94	1.92	1.89	1.86	1.83	1.79	1.75	1.70	1.66	1.62	1.57	1.52	1.47
	4.8	2.00	1.97	1.94	1.93	1.89	1.87	1.84	1.80	1.76	1.72	1.68	1.64	1.59	1.55	1.50
	5.0	2.00	1.97	1.94	1.93	1.90	1.87	1.85	1.81	1.77	1.73	1.69	1.65	1.61	1.57	1.53
(b) Upper k_L	1.8	1.65	...	...	...	...	...	...	...	...	...	...	...	...	...	...
	2.0	1.76	1.82	1.86	1.89	...	...	...	...	...	...	...	...	...	...	...
	2.2	1.83	1.89	1.93	1.96	2.00	2.04	2.06	...	...	...	...	...	...	...	...
	2.4	1.88	1.94	1.98	2.01	2.05	2.08	2.11	2.15	...	...	...	...	...	...	...
	2.6	1.92	1.97	2.01	2.03	2.08	2.11	2.14	2.18	2.22	...	...	...	...	...	...
	2.8	1.94	1.99	2.03	2.05	2.09	2.13	2.15	2.20	2.24	2.27	...	...	...	...	...
	3.0	1.96	2.01	2.04	2.06	2.10	2.13	2.16	2.21	2.25	2.28	2.32	...	...	...	...
	3.2	1.97	2.02	2.05	2.07	2.11	2.14	2.16	2.21	2.25	2.29	2.32	2.35	2.38	...	...
	3.4	1.98	2.02	2.05	2.07	2.11	2.14	2.16	2.21	2.25	2.28	2.32	2.35	2.38	2.41	...
	3.6	1.99	2.02	2.05	2.07	2.11	2.14	2.16	2.20	2.24	2.28	2.31	2.34	2.37	2.41	2.44
	3.8	1.99	2.03	2.05	2.07	2.11	2.13	2.16	2.20	2.24	2.27	2.30	2.33	2.36	2.40	2.43
	4.0	1.99	2.03	2.05	2.07	2.11	2.13	2.15	2.19	2.23	2.26	2.29	2.32	2.35	2.38	2.41
	4.2	2.00	2.03	2.05	2.07	2.10	2.13	2.15	2.19	2.22	2.25	2.28	2.31	2.34	2.37	2.40
	4.4	2.00	2.03	2.05	2.07	2.10	2.13	2.15	2.18	2.22	2.25	2.28	2.31	2.33	2.36	2.39
	4.6	2.00	2.03	2.05	2.07	2.10	2.12	2.14	2.18	2.21	2.24	2.27	2.30	2.32	2.35	2.38
	4.8	2.00	2.03	2.05	2.07	2.10	2.12	2.14	2.17	2.21	2.23	2.26	2.29	2.31	2.34	2.36
	5.0	2.00	2.03	2.05	2.07	2.09	92.1	22.1	2.17	2.20	2.23	2.25	42.2	2.30	82.3	2.35

NOTES—This table was reproduced from *Biometrika Tables for Statisticians*, Vol. 1, p. 207, with the kind permission of the Biometrika Trust. The Biometrika Tables also give the lower and upper 0.5, 1.0, and 5 percent points. Use for a large sample only, say $n \geq 250$. Take $\mu = \overline{X}$ and $\sigma = s$.
[a] When $g_1 > 0$, the skewness is taken to be positive, and the deviates for the lower percentage points are negative.

As a first step, one should scan the data to assess its approach to the Normal law. We suggest dividing g_1 and g_2 by their standard errors and if either ratio exceeds 3 then look to see if there is an outlier. An *outlier* is an observation so small or so large that there are no other observations near it and so extreme that persons familiar with the measurements can assert that such extreme value will not arise in the future under ordinary conditions. A glance at Fig. 2 suggests the presence of outliers but we must suppose that the second criterion was not satisfied.

If any observations seem to be outliers then discard them. If n is very large, say $n > 10000$, then use the percentile estimator based on the order statistics. If the ratios are both below 3 then use the Normal law for smaller sample sizes. If n is between 1000 and 10000 but the ratios suggest skewness and/or kurtosis, then use the cumulative frequency function. For smaller sample sizes and evidence of skewness and/or kurtosis, use the Pearson system curves. Obviously, these are rough guidelines and the user must adapt them to the actual situation by trying alternative calculations and then judging the most reasonable.

NOTE ON TOLERANCE LIMITS

In Sections 33 through 34, the percentages of X values estimated to be within a specified range pertain only to the given sample of data which is being represented succinctly by selected statistics, $\overline{X}$, s, etc. The Pearson curves used to derive these percentages are used simply as graduation formulas for the histogram of the sample data. The aim of Sections 33 to 34 is to indicate how much information about the sample is given by $\overline{X}$, s, g_1, and g_2. *It should be carefully noted that in an analysis of this kind the selected ranges of X and associated percentages are not to be confused with what in the statistical literature are called "tolerance limits."*

In statistical analysis, tolerance limits are values on the X scale that denote a range which may be stated to contain a specified minimum percentage of the values in the population there being attached to this statement a coefficient indicating the degree of confidence in its truth. For example, with reference to a random sample of 400 items, it may be said, with a 0.91 probability of being right, that 99 percent of the values in the population from which the sample came will be in the interval $X_{(400)} - X_{(1)}$ where $X_{(400)}$ and $X_{(1)}$

are respectively the largest and smallest values in the sample. If the population distribution is known to be Normal it might also be said, with a 0.90 probability of being right, that 99 percent of the values of the population will lie in the interval $\overline{X} \pm 2.703s$. Further information on statistical tolerances of this kind is presented in Refs. **6**, **7**, and **18**.

31. Use of Coefficient of Variation Instead of the Standard Deviation

So far as quantity of information is concerned, the presentation of the sample coefficient of variation, cv, together with the average, $\overline{X}$, is equivalent to presenting the sample standard deviation, s, and the average, $\overline{X}$, since s may be computed directly from the values of $cv = s/\overline{X}$ and $\overline{X}$. In fact, the sample coefficient of variation (multiplied by 100) is merely the sample standard deviation, s, expressed as a percentage of the average, $\overline{X}$. The coefficient of variation is sometimes useful in presentations whose purpose is to compare variabilities, relative to the averages, of two or more distributions. It is also called the relative standard deviation or relative error.

Example 1

Table 10 presents strength test results for two different materials. It can be seen that whereas the standard deviation for Material B is less than the standard deviation for Material A, the latter shows the greater relative variability as measured by the coefficient of variability.

TABLE 10. Strength test results.

MATERIAL	NUMBER OF OBSERVATIONS, n	AVERAGE STRENGTH, LB, $\overline{X}$	STANDARD DEVIATION, LB, s	COEFFICIENT OF VARIATION, % cv
A	160	1100	225	20.4
B	150	800	200	25.0

TABLE 11. Data for two test conditions.

TEST CONDITION	NUMBER OF SPECIMENS n	AVERAGE LIFE σ	STANDARD DEVIATION s	COEFFICIENT OF VARIATION, cv, %
A	50	14 h	4.2 h	30.0
B	50	80 min	23.2 min	29.0

The coefficient of variation is particularly applicable in reporting the results of certain measurements where the variability, σ, is known or suspected to depend upon the level of the measurements. Such a situation may be encountered when it is desired to compare the variability (*a*) of physical properties of related materials usually at different levels, (*b*) of the performance of a material under two different test conditions, or (*c*) of analyses for a specific element or compound present in different concentrations.

Example 2

The performance of a material may be tested under widely different test conditions as for instance in a standard life test and in an accelerated life test. Further, the units of measurement of the accelerated life tester may be in minutes and of the standard tester, in hours. The data shown in Table 11 indicate essentially the same relative variability of performance for the two test conditions.

32. General Comment on Observed Frequency Distributions of a Series of ASTM Observations

Experience with frequency distributions for physical characteristics of materials and manufactured products prompts the committee to insert a comment at this point. We have yet to find an observed frequency distribution of over 100 observations of a quality characteristic and purporting to represent essentially uniform conditions, that has less than 96 percent of its values within the range $\overline{X} \pm 3s$. For a Normal distribution, 99.7 percent of the cases should theoretically lie between $\mu \pm 3\sigma$ as indicated in Fig. 15.

Taking this as a starting point and considering the fact that in ASTM work the intention is, in general, to avoid throwing together into a single series data obtained under widely different conditions—different in an important sense in respect to the characteristic under inquiry—we believe that it is possible, in general, to use the methods indicated in Sections 33 and 34 for making rough estimates of the observed percentages of a frequency distribution, at least for making estimates (per Section 33) for symmetric ranges around the average, that is, $\overline{X} \pm ks$. This belief depends, to be sure, upon our own experience with frequency distributions and upon the observation that such distributions tend, in general, to be unimodal—to have a single peak—as in Fig. 14.

Discriminate use of these methods is, of course, presumed. The methods suggested for controlled conditions could not be expected to give satisfactory results if the parent distribution were one like that shown in Fig. 16—a bimodal distribution representing two different sets of conditions. Here, however, the methods could be applied separately to each of the two rational subgroups of data.

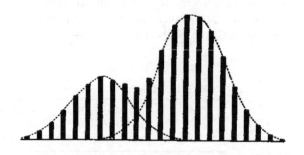

Fig. 16—A bimodal distribution arising from two different systems of causes.

33. Summary—Amount of Information Contained in Simple Functions of the Data

The material given in Sections 24 to 32, inclusive, may be summarized as follows:

1. If a sample of observations of a single variable is obtained under controlled conditions, much of the total information contained therein may be made available by presenting four functions—the average $\overline{X}$, the standard deviation, s, the skewness, g_1 the kurtosis, g_2 and the number n, of observations. Of the four functions, $\overline{X}$ and s contribute most; g_1 and g_2 contribute in accord with how small or how large are their standard errors, namely $\sqrt{6/n}$ and $\sqrt{24/n}$.

2. The average, $\overline{X}$, and the standard deviation, s, give some information even for data that are not obtained under controlled conditions.

3. No single function, such as the average, of a sample of observations is capable of giving much of the total information contained therein unless the sample is from a universe that is itself characterized by a single parameter. To be confident, the population that has this characteristic will usually require much previous experience with the kind of material or phenomenon under study.

Just what functions of the data should be presented, in any instance, depends on what uses are to be made of the data. This leads to a consideration of what constitutes the "essential information."

ESSENTIAL INFORMATION

34. Introduction

Presentation of data presumes some intended use either by others or by the author as supporting evidence for his or her conclusions. The objective is to present that portion of the total information given by the original data that is believed to be essential for the intended use. **Essential information** will be described as follows: "We take data to answer specific questions. We shall say that a set of statistics (functions) for a given set of data contains the **essential information** given by the data when, through the use of these statistics, we can answer the questions in such a way that further analysis of the data will not modify our answers to a practical extent," (taken from **PART 2**, Ref. **1**).

The **Preface** to this Manual lists some of the objectives of gathering ASTM data of the type under discussion—a sample of observations of a single variable. Each such sample constitutes an observed frequency distribution, and the information contained therein should be used efficiently in answering the questions that have been raised.

35. What Functions of the Data Contain the Essential Information

The nature of the questions asked determine what part of the total information in the data constitutes the essential information for use in interpretation.

If we are interested in the percentages of the total number of observations that have values above (or below) several values on the scale of measurement, the essential information may be contained in a tabular grouped frequency distribution plus a statement of the number of observations n. But even here, if n is large and if the data represent controlled conditions, the essential information may be contained in the four sample functions—the average $\overline{X}$, the standard deviation s, the skewness g_1, the kurtosis g_2, and the number of observations n. If we are interested in the average and variability of the quality of a material, or in the average quality of a material and some measure of the variability of averages for successive samples, or in a comparison of the average and variability of the quality of one material with that of other materials, or in the error of measurement of a test, or the like, then the essential information may be contained in the $\overline{X}$, s, and n of each sample of observations. Here, if n is small, say 10 or less, much of the essential information may be contained in the $\overline{X}$, R (range), and n of each sample of observations. The reason for use of R when n < 10 is as follows:

It is important to note (see Ref. **8**) that the expected value of the range R (largest observed value minus smallest observed value) for samples of n observations each, drawn from a Normal universe having a standard deviation σ varies with sample size in the following manner.

The expected value of the range is 2.1σ for $n = 4$, 3.1σ for $n = 10$, 3.9σ for $n = 25$, and 6.1σ for $n = 500$. From this it is seen that in sampling from a Normal population, the spread between the maximum and the minimum observation may be expected to be about twice as great for a sample of 25, and about three times as great for a sample of 500, as for a sample of 4. For this reason, n should *always* be given in presentations which give R. In general, it is better not to use R if n exceeds 12.

If we are also interested in the percentage of the total quantity of product that does not conform to specified limits, then part of the essential information may be contained in the observed value of fraction defective p. The conditions under which the data are obtained should always be indicated, i.e., (a) controlled, (b) uncontrolled, or (c) unknown.

If the conditions under which the data were obtained were not controlled, then the maximum and minimum observations may contain information of value.

It is to be carefully noted that if our interest goes beyond the sample data themselves to the processes that generated the samples or might generate similar samples in the future, we need to consider errors that may arise from sampling. The problems of sampling errors that arise in estimating process means, variances, and percentages are discussed in **PART 2**. For discussions of sampling errors in comparisons of means and variabilities of different samples, the reader is referred to texts on statistical theory (for example, Ref. **9**). The intention here is simply to note those statistics, those functions of the sample data, which would be useful in making such comparisons and consequently should be reported in the presentation of sample data.

36. Presenting $\overline{X}$ Only Versus Presenting $\overline{X}$ and s

Presentation of the essential information contained in a sample of observations commonly consists in presenting $\overline{X}, s$, and n. Sometimes the average alone is given—no record is made of the dispersion of the observed values nor of the number of observations taken. For example, Table 12 gives the observed average tensile strength for several materials under several conditions.

TABLE 12. Information of value may be lost if only the average is presented.

	Tensile Strength psi		
Material	Condition a Average, $\overline{x}$	Condition b Average, $\overline{x}$	Condition c Average, $\overline{x}$
A	51 430	47 200	49 010
B	59 060	57 380	60 700
C	57 710	74 920	80 460

The objective quality in each instance is a frequency distribution, from which the set of observed values might be considered as a sample. Presenting merely the average, and failing to present some measure of dispersion

TABLE 13. Presentation of essential information (data of **TABLE** 8).

| | | Tensile Strength, psi | | | | | |
| | | Condition a | | Condition b | | Condition c | |
Material	Tests	Average, $\overline{X}$	Standard Deviation, s	Average, $\overline{X}$	Standard Deviation, s	Average, $\overline{X}$	Standard Deviation, s
A	20	51 430	920	47 200	830	49 010	1070
B	18	59 060	1320	57 380	1 360	60 700	1480
C	27	75 710	1840	74 920	1 650	80 460	1910

and the number of observations generally loses much information of value. Table 13 corresponds to Table 12 and provides what will usually be considered as the essential information for several sets of observations, such as data collected in investigations conducted for the purpose of comparing the quality of different materials.

37. Observed Relationships

ASTM work often requires the presentation of data showing the observed relationship between two variables. Although this subject does not fall strictly within the scope of **PART 1** of the Manual, the following material is included for general information. Attention will be given here to one type of relationship, where one of the two variables is of the nature of temperature or time—one that is controlled at will by the investigator and considered for all practical purposes as capable of "exact" measurement, free from experimental errors. (The problem of presenting information on the observed relationship between *two statistical*

variables, such as hardness and tensile strength of an alloy sheet material, is more complex and will not be treated here. For further information, see Refs. **1**, **2**, and **9**.) Such relationships are commonly presented in the form of a chart consisting of a series of plotted points and straight lines connecting the points or a smooth curve which has been "fitted" to the points by some method or other. This section will consider merely the information associated with the plotted points, i.e., scatter diagrams.

Figure 17 gives an example of such an observed relationship. (Data are from records of shelf life tests on die-cast metals and alloys, former Subcommittee 15 of ASTM Committee B-2 on Non-Ferrous Metals and Alloys.) At each successive stage of an investigation to determine the effect of aging on several alloys, five specimens of each alloy were tested for tensile strength by each of several laboratories. The curve shows the results obtained by one laboratory for one of these alloys. Each of the plotted points is the average of five observed values of tensile strength and thus attempts to summarize an observed frequency distribution.

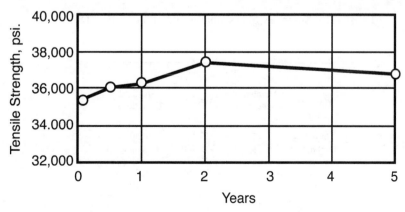

Fig. 17—**Example of graph showing an observed relationship.**

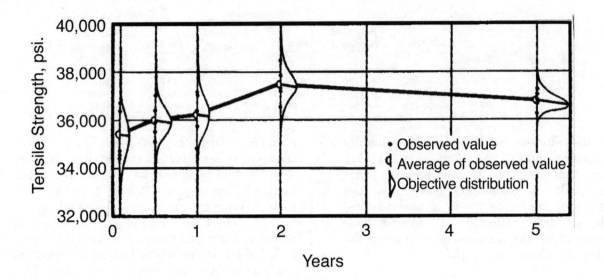

Fig. 18—Showing pictorially what lies back of the plotted points in Fig. 17. Each plotted point in Fig. 17 is the average of a sample from a universe of possible observations.

Figure 18 has been drawn to show pictorially what is behind the scenes. The five observations made at each stage of the life history of the alloy constitute a sample from a universe of possible values of tensile strength—an objective frequency distribution whose spread is dependent on the inherent variability of the tensile strength of the alloy and on the error of testing. The dots represent the observed values of tensile strength and the bell-shaped curves the objective distributions. In such instances, the essential information contained in the data may be made available by supplementing the graph by a tabulation of the averages, the standard deviations, and the number of observations for the plotted points in the manner shown in Table 14.

TABLE 14. Summary of essential information for Fig. 18.

Time of Test	Number of Specimens	Tensile strength, psi Average, $\bar{x}$	Standard Deviation, s
Initial	5	35 400	950
6 months	5	35 980	668
1 year	5	36 220	869
2 years	5	37 460	655
5 years	5	36 800	319

38. Summary: Essential Information

The material given in Sections 34 to 37, inclusive, may be summarized as follows.

1. What constitutes the *essential information* in any particular instance depends on the nature of the questions to be answered, and on the nature of the hypotheses that we are willing to make based on available information. Even when measurements of a quality characteristic are made under the same essential conditions, the objective quality is a *frequency distribution* that cannot be adequately described by any single numerical value.

2. Given a series of observations of a single variable arising from the same essential conditions, it is the opinion of the committee that the average, $\overline{X}$, the standard deviation, s, and the number, n, of observations contain the essential information for a majority of the uses made of such data in ASTM work.

NOTE

If the observations are not obtained under the same essential conditions, analysis and presentation by the control chart method, in which *order* (see **PART 3** of this Manual) is taken into account by rational subgrouping of observations, commonly provides important additional information.

PRESENTATION OF RELEVANT INFORMATION

39. Introduction

Empirical knowledge is not contained in the observed data alone, rather it arises from interpretation—an act of thought. (For an important discussion on the significance of prior information and hypothesis in the interpretation of data, see Ref. **10**; a treatise on the philosophy of probable inference which is of basic importance in the interpretation of any and all data is presented in Ref. **11**.) Interpretation consists in testing hypotheses based on prior knowledge. Data constitute but a part of the information used in interpretation—the judgments that are made depend as well on pertinent collateral information, much of which may be of a qualitative rather than of a quantitative nature.

If the data are to furnish a basis for most valid prediction, they must be obtained under controlled conditions and must be free from constant errors of measurement. Mere presentation does not alter the goodness or badness of data. However, the usefulness of good data may be enhanced by the manner in which they are presented.

40. Relevant Information

Presented data should be accompanied by any or all available relevant information, particularly information on *precisely* the field within which the measurements are supposed to hold and the condition under which they were made, and evidence that the data are good. Among the specific things that may be presented with ASTM data to assist others in interpreting them or to build up confidence in the interpretation made by an author are:

1. The kind, grade, and character of material or product tested.

2. The mode and conditions of production, if this has a bearing on the feature under inquiry.

3. The method of selecting the sample; steps taken to ensure its randomness or representativeness. (The manner in which the sample is taken has an important bearing on the interpretability of data and is discussed by Dodge (see Ref. **12**).)

4. The specific method of test (if an ASTM or other standard test, so state; together with any modifications of procedure).

5. The specific conditions of test, particularly the regulation of factors that are known to have an influence on the feature under inquiry.

6. The precautions or steps taken to eliminate systematic or constant errors of observation.

7. The difficulties encountered and eliminated during the investigation.

8. Information regarding parallel but independent paths of approach to the end results.

9. Evidence that the data were obtained under controlled conditions; the results of statistical tests made to support belief in the constancy of conditions, in respect to the physical tests made or the material tested, or both. (Here, we mean constancy in the statistical sense, which encompasses the thought of stability of conditions from one time to another and from one place to another. This state of affairs is commonly referred to as "statistical control." Statistical criteria have been developed by means of which we may judge when controlled conditions exist. Their character and mode of application are given in **PART 3** of this Manual; see also Ref. **13**.)

Much of this information may be qualitative in character, some may even be vague, yet without it the interpretation of the data and the conclusions reached may be misleading or of little value to others.

41. Evidence of Control

One of the fundamental requirements of good data is that they should be obtained under controlled conditions. The interpretation of the observed results of an investigation depends on whether or not there is justification for believing that the conditions were controlled.

If the data are numerous and statistical tests for control are made, evidence of control may be presented by giving the results of these tests. (For examples, see Refs. **14–18**.) Such quantitative evidence greatly strengthens inductive arguments. In any case, it is important to indicate clearly just what precautions were taken to control the essential conditions. Without tangible evidence of this character, the reader's degree of rational belief in the results presented will depend on his faith in the ability of the investigator to eliminate all causes of lack of constancy.

RECOMMENDATIONS

42. Recommendations for Presentation of Data

The following recommendations for presentation of data apply for the case where one has at hand a sample of n observations of a single variable obtained under the same essential conditions.

1. Present as a minimum, the average, the standard deviation, and the number of observations. *Always* state the number of observations taken.

2. If the number of observations is moderately large ($n > 30$), present also

the value of the skewness, g_1, and the value of the kurtosis, g_2. An additional procedure when n is large ($n>100$) is to present a graphical representation, such as a grouped frequency distribution.

3. If the data were not obtained under controlled conditions and it is desired to give information regarding the extreme observed effects of assignable causes, present the values of the maximum and minimum observations in addition to the average, the standard deviation, and the number of observations.

4. Present as much evidence as possible that the data were obtained under controlled conditions.

5. Present relevant information on precisely (a) the field within which the measurements are believed valid and (b) the conditions under which they were made.

REFERENCES

[1] Shewhart, W. A., *Economic Control of Quality of Manufactured Product*, Van Nostrand, New York, 1931; republished by ASQC Quality Press, Milwaukee, WI, 1980.

[2] Yule, G. U. and Kendall, M. G., *An Introduction to the Theory of Statistics*, 14th ed., Charles Griffin and Company, Ltd., London, 1950.

[3] Box, G. E. P., Hunter, W. G., and Hunter, J. S., *Statistics for Experimenters*, Wiley, New York, 1978, pp. 329–330.

[4] Tukey, J. W., *Exploratory Data Analysis*, Addison-Wesley, Reading, PA, 1977, pp. 1–26.

[5] Elderton, W. P. and Johnson, N. L., *Systems of Frequency Curves*, Cambridge University Press, Bentley House, London, 1969.

[6] Duncan, A. J., *Quality Control and Industrial Statistics*, 5th ed., Chapter 6, Sections 4 and 5, Richard D. Irwin, Inc., Home Wood, IL, 1986.

[7] Bowker, A. H. and Lieberman, G. J., *Engineering Statistics*, 2nd ed., Section 8.12, Prentice-Hall, 1972.

[8] Tippett, L. H. C., "On the Extreme Individuals and the Range of Samples Taken From a Normal Population," *Biometrika*, Vol. 17, Dec. 1925, pp. 364–387.

[9] Hoel, P. G., *Introduction to Mathematical Statistics*, 5th ed., Wiley, New York, 1984.

[10] Lewis, C. I., *Mind and the World Order*, Scribner, New York, 1929.

[11] Keynes, J. M., *A Treatise on Probability*, MacMillan, New York, 1921.

[12] Dodge, H. F., "Statistical Control in Sampling Inspection," presented at a round table discussion on "Acquisition of Good Data," held at the 1932 Annual Meeting of the American Society for Testing and Materials; published in *American Machinist*, 26 Oct. and 9 Nov. 1932.

[13] Pearson, E. S., "A Survey of the Uses of Statistical Method in the Control and Standardization of the Quality of Manufactured Products," *Journal of the Royal Statistical Society*, Vol. XCVI, Part 11, 1933, pp. 21–60.

[14] Passano, R. F., "Controlled Data from an Immersion Test," *Proceedings*, American Society for Testing and Materials, Vol. 32, Part 2, 1932, p. 468.

[15] Skinker, M. F., "Application of Control Analysis to the Quality of Varnished Cambric Tape," *Proceedings*, American Society for Testing and Materials, Vol. 32, Part 3, 1932, p. 670.

[16] Passano, R. F. and Nagley, F. R., "Consistent Data Showing the Influences of Water Velocity and Time on the Corrosion of Iron," *Proceedings*, American Society for Testing and Materials, Vol. 33, Part 2, p. 387.

[17] Chancellor, W. C., "Application of Statistical Methods to the Solution of Metallurgical Problems in Steel Plant," *Proceedings*, American Society for Testing and Materials, Vol. 34, Part 2, 1934, p. 891.

[18] Ott, E. R., Schilling, E. G., and Neubauer, D. V., *Process Quality Control*, 3rd ed., McGraw-Hill, New York, N.Y., 2000.

2

Presenting Plus or Minus Limits of Uncertainty of an Observed Average

GLOSSARY OF SYMBOLS USED IN PART 2

μ	Population mean
a	Factor, given in Table 2 of **PART 2,** for computing confidence limits for μ associated with a desired value of probability, $P,$ and a given number of observations, n
k	Deviation of a Normal variable
n	Number of observed values (observations)
p	Sample fraction nonconforming
p'	Population fraction nonconforming
σ	Population standard deviation
P	Probability; used in **PART 2** to designate the probability associated with confidence limits; relative frequency with which the averages μ of sampled populations may be expected to be included within the confidence limits (for μ) computed from samples
s	Sample standard deviation
$\hat{\sigma}$	Estimate of σ based on several samples
X	Observed value of a measurable characteristic; specific observed values are designated $X_1, X_2, X_3,$ etc.; also used to designate a measurable characteristic
$\overline{X}$	*Sample average* (arithmetic mean), the sum of the n observed values in a set divided by n

1. Purpose

PART 2 of the Manual discusses the problem of presenting plus or minus limits to indicate the uncertainty of the average of a number of observations obtained under the same essential conditions, and suggests a form of presentation for use in ASTM reports and publications where needed.

2. The Problem

An observed average $\overline{X}$, is subject to the uncertainties that arise from sampling fluctuations and tends to differ from the population mean. The smaller the number of observations, n, the larger the fluctuations are likely to be.

With a set of n observed values of a variable X, whose average (arithmetic mean) is $\overline{X}$, as in Table 1, it is often desired to interpret the results in some way. One way is to construct an interval such that the mean, $\mu = 573.2 \pm 3.5$ lb., lies within limits being established from the quantitative data along with the implications that the mean μ of the population sampled is included within these limits with a specified probability. How should such limits be computed, and what meaning may be attached to them?

TABLE 1. Breaking strength of ten specimens of 0.104-in. hard-drawn copper wire.

Specimen	Breaking Strength, X, lb
1	578
2	572
3	570
4	568
5	572
6	570
7	570
8	572
9	576
10	584
$n = 10$	5732
Average, $\overline{X}$	573.2
Standard deviation, s	4.83

NOTE

The mean μ is the value of $\overline{X}$ that would be approached as a statistical limit as more and more observations were obtained under the same essential conditions and their cumulative averages computed.

3. Theoretical Background

Mention should be made of the practice, now mostly out of date in scientific work, of recording such limits as

$$\overline{X} \pm 0.6745 \frac{s}{\sqrt{n}}$$

where

$\overline{X}$ = observed average,
s = observed standard deviation, and
n = number of observations,

and referring to the value $0.6745\,s/\sqrt{n}$ as the "probable error" of the observed average, $\overline{X}$. (Here the value of 0.6745 corresponds to the Normal law probability of 0.50, see Table 8 of **PART 1**.) The term "probable error" and the probability value of 0.50 properly apply to the errors of sampling when sampling from a universe whose average, μ, and whose standard deviation, σ, are *known* (these terms apply to limits $\mu \pm 0.6745\,\sigma/\sqrt{n}$), but they do not apply in the inverse problem when merely sample values of $\overline{X}$ and s are given.

Investigation of this problem [1–3] has given a more satisfactory alternative (section 4), a procedure which provides limits that have a definite operational meaning.

NOTE

While the method of Section 4 represents the best that can be done at present in interpreting a sample $\overline{X}$ and s, when no other information regarding the variability of the population is available, a much more satisfactory interpretation can be made in general if other information regarding the variability of the population is at hand, such as a series of samples from the universe or similar populations for each of which a value of s or R is computed. If s or R displays statistical control, as outlined in **PART 3** of this Manual, and a sufficient number of samples (preferably 20 or more) are available to obtain a reasonably precise estimate of σ, designated as $\hat{\sigma}$, the limits of uncertainty for a sample containing any number of observations, n, and arising from a population whose true standard deviation can be presumed equal to $\hat{\sigma}$, can be computed from the following formula:

$$\overline{X} \pm k\frac{\hat{\sigma}}{\sqrt{n}}$$

where $k = 1.645$, 1.960, and 2.576 for probabilities of $P = 0.90$, 0.95, and 0.99, respectively.

4. Computation of Limits

The following procedure applies to any long run *series* of problems for *each* of which the following conditions are met:

Given: A sample of n observations, X_1, X_2, X_3, ..., X_n, having an average = $\overline{X}$ and a standard deviation = s.

TABLE 2. Factors for calculating 90, 95, and 99 percent confidence limits for averages.[a]

Number of Observations in Sample, n	Confidence Limits, * $\overline{X} \pm as$		
	90% Confidence Limits $(P = 0.90)$ Value of a	95% Confidence Limits $(P = 0.95)$ Value of a	99% Confidence Limits $(P = 0.99)$ Value of a
4	1.177	1.591	2.921
5	0.953	1.241	2.059
6	0.823	1.050	1.646
7	0.734	0.925	1.401
8	0.670	0.836	1.237
9	0.620	0.769	1.118
10	0.580	0.715	1.028
11	0.546	0.672	0.955
12	0.518	0.635	0.897
13	0.494	0.604	0.847
14	0.473	0.577	0.805
15	0.455	0.554	0.769
16	0.438	0.533	0.737
17	0.423	0.514	0.708
18	0.410	0.497	0.683
19	0.398	0.482	0.660
20	0.387	0.468	0.640
21	0.376	0.455	0.621
22	0.367	0.443	0.604
23	0.358	0.432	0.588
24	0.350	0.422	0.573
25	0.342	0.413	0.559
n greater than 25	$a = \dfrac{1.645}{\sqrt{n}}$ approximately	$a = \dfrac{1.960}{\sqrt{n}}$ approximately	$a = \dfrac{2.576}{\sqrt{n}}$ approximately

*Limits which may be expected to include μ (9 times in 10, 95 times in 100, or 99 times in 100) in a series of problems, each involving a single sample of n observations.

Values of a computed from Fisher, R. A., "Table of t," *Statistical Methods for Research Workers,* Table IV based on Student's distribution of t.

[a] Recomputed in 1975. The a of this Table 2 equals Fisher's t for $n - 1$ degrees of freedom divided by $\sqrt{n}$. See also Fig. 1.

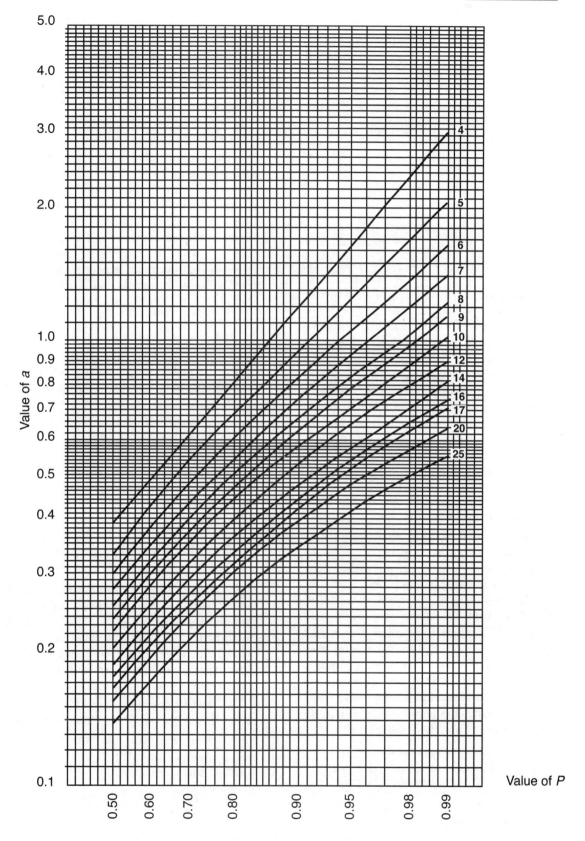

FIG. 1—Curves giving factors for calculating 50 to 99 percent confidence limits for averages (see also Table 2). Redrawn in 1975 for new values of *a*. Error in reading *a* not likely to be > 0.01. The numbers printed by curves are the sample sizes *n*.

Conditions: (*a*) The population sampled is homogeneous (statistically controlled) in respect to *X*, the variable measured. (*b*) The distribution of *X* for the population sampled is approximately Normal. (*c*) The sample is a random sample.[1]

Procedure: Compute limits

$$\overline{X} \pm as$$

where the value of *a* is given in Table 2 for three values of *P* and for various values of *n*.

Meaning: If the values of *a* given in Table 2 for *P* = 0.95 are used in a series of such problems, then, in the long run, we may expect 95 percent of the intervals bounded by the limits so computed, to include the population averages μ of the populations sampled. If in each instance, we were to assert that μ is included within the limits computed, we should expect to be correct 95 times in 100 and in error 5 times in 100; that is, the statement "μ is included within the interval so computed" has a probability of 0.95 of being correct. But there would be no operational meaning in the following statement made in any one instance: "The probability is 0.95 that μ falls within the limits computed in this case" since μ either does or does not fall within the limits. It should also be emphasized that even in repeated sampling from the *same* population, the interval defined by the limits $\overline{X} \pm as$ will vary in width and position from sample to sample, particularly with small samples (see Fig. 2). It is this series of ranges fluctuating in size and position which will include, ideally, the population mean μ, 95 times out of 100 for *P* = 0.95.

These limits are commonly referred to as "confidence limits" [4,5,18] for the three columns of Table 2 they may be referred to as the "90 percent confidence limits," "95 percent confidence limits" and "99 percent confidence limits," respectively.

The magnitude *P* = 0.95 applies to the *series of samples*, and is approached as a statistical limit as the number of instances in the series is increased indefinitely; hence it signifies "statistical probability." If the values of *a* given in Table 2 for *P* = 0.99 are used in a series of samples, we may, in like manner, expect 99 percent of the sample intervals so computed to include the population mean μ.

Other values of *P* could, of course, be used if desired—the use of chances of 95 in 100, or 99 in 100 are, however, often found convenient in engineering presentations. Approximate values of *a* for other values of *P* may be read from the curves in Fig. 1, for samples of *n* = 25 or less.

For larger samples (*n* greater than 25), the constants, 1.645, 1.960, and 2.576, in the expressions

$$a = \frac{1.645}{\sqrt{n}}, \quad a = \frac{1.960}{\sqrt{n}}, \text{ and } a = \frac{2.576}{\sqrt{n}}$$

at the foot of Table 2 are obtained directly from Normal law integral tables for probability values of 0.90, 0.95, and 0.99. To find the value of this constant for any other value of *P*, consult any standard text on statistical methods or read the value approximately on the "*k*" scale of Fig. 15 of **PART 1** of this Manual. For example, use of $a = 1/\sqrt{n}$ yields P = 68.27% and the limits plus or minus one standard error, which some scientific journals print without noting a percent.

---

[1] If the population sampled is finite, that is, made up of a finite number of separate units that may be measured in respect to the variable, *X*, and if interest centers on the μ of this population, then this procedure assumes that the number of units, *n*, in the sample is relatively small compared with the number of units, *N*, in the population, say *n* is less than about 5 percent of *N*. However, correction for relative size of sample can be made by multiplying *s* by the factor $\sqrt{1 - (n/N)}$. On the other hand, if interest centers on the μ of the underlying process or source of the finite population, then this correction factor is not used.

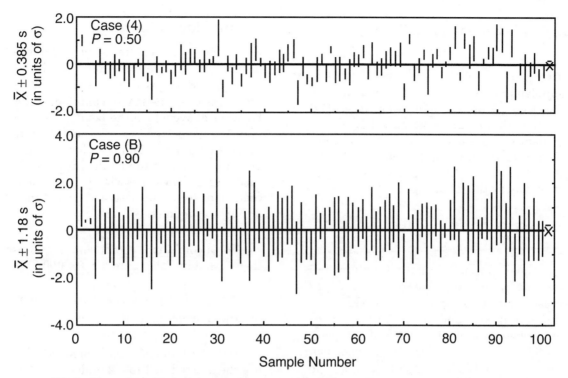

FIG. 2—Illustration showing computed intervals based on sampling experiments; 100 samples of *n* = 4 observations each, from a Normal universe having μ = 0 and σ = case A is taken from Fig. 8 of Ref. 2, and Case B gives corresponding intervals for limits $\overline{X} \pm 1.18\,s$, **based on P = 0.90.**

5. Experimental Illustration

Figure 2 gives two diagrams illustrating the results of sampling experiments for samples of *n* = 4 observations each drawn from a Normal population, for values of Case A, *P* = 0.50, and Case B, *P* = 0.90. For Case A, the intervals for 51 out of 100 samples included μ and for Case B, 90 out of 100 included μ. If, in each instance (that is, for each sample) we had concluded that the population mean μ is included within the limits shown for Case A, we would have been correct 51 times and in error 49 times, which is a reasonable variation from the expectancy of being correct 50 percent of the time.

In this experiment all samples were taken from the same population. However, the same reasoning applies to a series of samples each drawn from a population from the same universe as evidenced by conformance to the three conditions set forth in Section 4.

6. Presentation of Data

In presentation of data, if it is desired to give limits of this kind, it is quite important that the probability associated with the limits be clearly indicated. The three values *P* = 0.90, *P* = 0.95, and *P* = 0.99 given in Table 2 (chances of 9 in 10, 95 in 100, and 99 in 100) are arbitrary choices that may be found convenient in practice.

Example

Given a sample of 10 observations of breaking strength of hard-drawn copper wire as in Table 1, for which

$$\overline{X} = 573.2 \text{ lb}$$
$$s = 4.83 \text{ lb}$$

Using this sample to define limits of uncertainty based on *P* = 0.95 (Table 2), we have

$$\overline{X} \pm 0.715s = 573.2 \pm 3.5$$
$$= 569.7 \text{ and } 576.7$$

Two pieces of information are needed to supplement this numerical result: (a) the fact that 95 in 100 limits were used, and (b) that this result is based solely on the evidence contained in 10 observations. Hence, in the presentation of such limits, it is desirable to give the results in some such way as the following

$$573.2 \pm 3.5 \text{ lb } (P = 0.95, \ n = 10)$$

The essential information contained in the data is, of course, covered by presenting $\overline{X}$, s, and n (see **PART 1** of this Manual) and the limits under discussion could be derived directly therefrom. If it is desired to present such limits in addition to $\overline{X}$, s, and n, the tabular arrangement given in Table 3 is suggested.

A satisfactory alternative is to give the $\pm$ value in the column designated "Average, $\overline{X}$" and to add a note giving the significance of this entry, as shown in Table 4. If one omits the note, it will be assumed that $a = 1/\sqrt{n}$ was used and that P = 68%.

7. One-Sided Limits

Sometimes we are interested in limits of uncertainty in one direction only. In this case we would present $\overline{X} + as$ or $\overline{X} - as$ (not both), a one-sided confidence limit below or above which the population mean may be expected to lie in a stated proportion of an indefinitely large number of samples. The a to use in this one-sided case and the associated confidence coefficient would be obtained from Table 2 or Fig. 1 as follows.

For a confidence coefficient of 0.95 use the a listed in Table 2 under $P = 0.90$.

For a confidence coefficient of 0.975 use the a listed in Table 2 under $P = 0.95$.

For a confidence coefficient of 0.995 use the a listed in Table 2 under $P = 0.99$.

In general, for a confidence coefficient of P_1 use the a derived from Fig. 1 for $P = 1 - 2(1 - P_1)$. For example, with $n = 10$, $\overline{X} = 573.2$, and $s = 4.83$ the one-sided upper $P_1 = 0.95$ confidence limit would be to use $a = 0.58$ for $P = 0.90$ in Table 2, which yields $573.2 + 0.58(4.83) = 573.2 + 2.8 = 576.0$.

TABLE 3. Suggested tabular arrangement.

NUMBER OF TESTS, n	AVERAGE, $\overline{X}$	LIMITS FOR μ (95 PERCENT CONFIDENCE LIMITS)	STANDARD DEVIATION, s
10	573.2	573.2 ± 3.5	4.83

TABLE 4. Alternative to Table 3.

NUMBER OF TESTS, n	AVERAGE, $\overline{X}$ [a]	STANDARD DEVIATION, s
10	573.2 ($\pm$ 3.5)	4.83

[a]The $\pm$ entry indicates 95 percent confidence limits for μ.

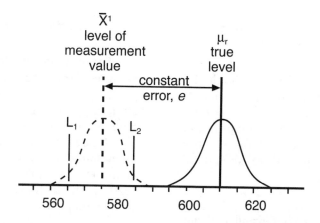

FIG. 3—Showing how plus or minus limits (L_1 and L_2) are unrelated to a systematic or constant error.

8. General Comments on the Use of Confidence Limits

In making use of limits of uncertainty of the type covered in this part, the engineer should keep in mind: (1) the restrictions as to (a) controlled conditions, (b) approximate Normality of population, (c) randomness of sample; and (2) the fact that the variability under consideration relates to fluctuations around the level of measurement values, whatever that may be, quite regardless of whether or not the population mean μ, of the measurement values is widely displaced from the true value, $μ_T$, of the thing measured, as a result of the systematic or *constant* errors present throughout the measurements.

For example, breaking strength values might center around a value of 575.0 lb (the population mean μ of the measurement values) with a scatter of individual observations represented by the dotted distribution curve of Fig. 3, whereas the true average $μ_T$ for the batch of wire under test is actually 610.0 lb, the difference between 575.0 and 610.0 representing a constant or systematic error present in *all* the observations as a result, say, of an incorrect adjustment of the testing machine.

The limits thus have meaning for series of like measurements, made under like conditions, *including* the same constant errors if any be present.

In the practical use of these limits, the engineer may not have assurance that conditions (a), (b), and (c) given in section 4 are met, hence it is not advisable to lay great emphasis on the exact magnitudes of the probabilities given in Table 2, but rather to consider them as orders of magnitude to be used as general guides.

9. Number of Places to be Retained in Computation and Presentation

The following working rule is recommended in carrying out computations incident to determining averages, standard deviations, and "limits for averages" of the kind here considered, for a sample of n observed values of a variable quantity:

"In all operations on the sample of n observed values, such as adding, subtracting, multiplying, dividing, squaring, extracting square root, etc., retain the equivalent of two more places of figures than in the single observed values. For example, if observed values are read or determined to the nearest 1 lb., carry numbers to the nearest 0.01 lb. in the computations; if observed values are read or determined to the nearest 10 lb., carry numbers to the nearest 0.1 lb. in the computations, etc."

Deleting places of figures should be done after computations are completed, in order to keep the final results substantially free from computation errors. In deleting places of figures the actual rounding-off procedure should be carried out as follows.[2]

1. When the figure next beyond the last figure or place to be retained is less than 5, the figure in the last place retained should be kept unchanged.

2. When the figure next beyond the last figure or place to be retained is more than 5, the figure in the last place retained should be increased by 1.

3. When the figure next beyond the last figure or place to be retained is 5, and (a) there are no figures, or only zeros, beyond this 5, if the figure in the last place to be retained is odd, it should be increased by 1; if even, it should be kept unchanged; but (b) if the 5 next beyond the figure in the last place to be retained is followed by any figures other than zero, the figure in the last place retained should be increased by 1, whether odd or even. For example, if in the following numbers, the places of figures in parenthesis are to be rejected

39 4(49) becomes 39 400,

39 4(50) becomes 39 400,

39 4(51) becomes 39 500, and

---

[2] This rounding-off procedure agrees with that adopted in ASTM Recommended Practice for Indicating which Places of Figures Are to be Considered Significant in Specified Limiting Values (E 29).

39 5(50) becomes 39 600.

The number of places of figures to be retained in presentation depends on what use is to be made of the results and on the sampling variation present. No general rule, therefore, can safely be laid down. The following working rule has, however, been found generally satisfactory by the committee in presenting the results of testing in technical investigations and development work:

(a) See **TABLE 5** for averages.

(b) For standard deviations, retain three places of figures.

(c) If "limits for averages" of the kind here considered are presented, retain the same places of figures as are retained for the average.

For example, if $n = 10$, and if observed values were obtained to the nearest 1 lb, present averages and "limits for averages" to the nearest 0.1 lb, and present the standard deviation to three places of figures. This is illustrated in the tabular presentation in section 6.

Rule (a) will result generally in one and conceivably in two doubtful places of figures in the average, that is, places which may have been affected by the rounding-off (or observation) of the n individual values to the nearest number of units stated in the first column of the table. Referring to Tables 3 and 4 the third place figures in the average, $\overline{X} = 573.2$, corresponding to the first place of figures in the ± 3.5 value is doubtful in this sense. One might conclude that it would be suitable to present the average to the nearest pound, thus,

$$573 \pm 3 \text{ lb } (P = 0.95, n = 10)$$

TABLE 5. Averages.

WHEN THE SINGLE VALUES ARE OBTAINED TO THE NEAREST	AND WHEN THE NUMBER OF OBSERVED VALUES IS		
0.1, 1, 10, etc., units	...	2 to 20	21 to 200
0.2, 2, 20, etc., units	less than 4	4 to 40	41 to 400
0.5, 5, 50, etc., units	less than 10	10 to 100	101 to 1000
RETAIN THE FOLLOWING NUMBER OF PLACES OF FIGURES IN THE AVERAGE	same number of places as in single values	1 more place than in single values	2 more places than in single values

TABLE 6. Effect of rounding.

	NOT ROUNDED			ROUNDED		
	LIMITS		DIFFERENCE		LIMITS	DIFFERENCE
573.5 ± 1.4	572.1	574.9	2.8	574 ± 1	573 575	2
573.5 ± 1.5	572.0	575.0	3.0	574 ± 2	572 576	4

This might be satisfactory for some purposes. However, the effect of such rounding off to the first place of figures of the plus or minus value may be quite pronounced if the first digit of the plus or minus value is small, as indicated in Table 6. If further use were to be made of these data—collecting additional observations to be combined with these, gathering other data to be compared with these, etc.— then the effect of such rounding off of $\overline{X}$ in a presentation might seriously interfere with proper subsequent use of the information.

The number of places of figures to be retained or to be used as a basis for action in specific cases cannot readily be made subject to any general rule. It is, therefore, recommended that in such cases the number of places be settled by definite agreements between the individuals or parties involved. In reports covering the acceptance and rejection of material ASTM E 29 gives specific rules that are applicable when reference is made to this recommended practice.

SUPPLEMENT A

Presenting Plus or Minus Limits of Uncertainty for σ —*Normal Distribution*[3]

When observations X_1, X_2, ..., X_n are made under controlled conditions, and there is reason to believe the distribution of X is Normal, two-sided confidence limits for the standard deviation of the population with confidence coefficient, P, will be given by lower confidence limit for

$$\hat{\sigma}_L = s\sqrt{(n-1)/\chi^2_{(1-P)/2}} \qquad (1)$$

upper confidence limit for

$$\hat{\sigma}_U = s\sqrt{(n-1)/\chi^2_{(1+P)/2}}$$

where the quantity $\chi^2_{(1-P)/2}$ (or $\chi^2_{(1+P)/2}$) is the χ^2 value of a chi-square variable with n

---

[3] The analysis is strictly valid only for an unlimited population such as presented by a manufacturing or measurement process. When the population sampled is relatively small compared with the sample size n, the reader is advised to consult a statistician.

– 1 degrees of freedom which is exceeded with probability $(1 - P)/2$ or $(1 + P)/2$ as found in most statistics text books.

To facilitate computation, Table 7 gives values of

$$b_{\mathrm{L}} = \sqrt{(n-1)/X^2_{(1-P)/2}} \quad \text{and}$$
$$b_{\mathrm{U}} = \sqrt{(n-1)/X^2_{(1+P)/2}} \qquad (2)$$

for $P = 0.90$, 0.95, and 0.99. Thus we have, for a Normal distribution, the estimate of the lower confidence limit for σ as

$$\hat{\sigma}_L = b_L s$$

and for the upper confidence limit $\qquad (3)$

$$\hat{\sigma}_U = b_U s$$

Example

Table 1 of **PART 2** gives the standard deviation of a sample of 10 observations of breaking strength of copper wire as $s = 4.83$ lb. If we assume that the breaking strength has a Normal distribution, which may actually be somewhat questionable, we have as 0.95 confidence limits for the universe standard deviation σ that yield a

TABLE 7. *b*-factors for calculating confidence limits for σ, Normal distribution.[a]

Number of Observations in Sample, n	90% Confidence Limits		95% Confidence Limits		99% Confidence Limits	
	b_{L}	b_{U}	b_{L}	b_{U}	b_{L}	b_{U}
2	0.510	16.0	0.446	31.9	0.356	159.5
3	0.578	4.41	0.521	6.29	0.434	14.1
4	0.619	2.92	0.566	3.73	0.484	6.47
5	0.649	2.37	0.600	2.87	0.518	4.40
6	0.671	2.09	0.625	2.45	0.547	3.48
7	0.690	1.91	0.645	2.20	0.569	2.98
8	0.705	1.80	0.661	2.04	0.587	2.66
9	0.718	1.71	0.676	1.92	0.603	2.44
10	0.730	1.64	0.688	1.83	0.618	2.28
11	0.739	1.59	0.698	1.75	0.630	2.15
12	0.747	1.55	0.709	1.70	0.641	2.06
13	0.756	1.51	0.718	1.65	0.651	1.98
14	0.762	1.49	0.725	1.61	0.660	1.91
15	0.769	1.46	0.732	1.58	0.669	1.85
16	0.775	1.44	0.739	1.55	0.676	1.81
17	0.780	1.42	0.745	1.52	0.683	1.76
18	0.785	1.40	0.750	1.50	0.690	1.73
19	0.789	1.38	0.756	1.48	0.696	1.70
20	0.794	1.37	0.760	1.46	0.702	1.67
21	0.798	1.35	0.765	1.44	0.707	1.64
22	0.801	1.35	0.769	1.43	0.712	1.62
23	0.806	1.34	0.773	1.41	0.717	1.60
24	0.808	1.33	0.777	1.40	0.721	1.58
25	0.812	1.32	0.780	1.39	0.725	1.56
26	0.814	1.31	0.785	1.38	0.730	1.54
27	0.818	1.30	0.788	1.37	0.734	1.52
28	0.821	1.29	0.791	1.36	0.738	1.51
29	0.823	1.29	0.793	1.35	0.741	1.50
30	0.825	1.28	0.797	1.35	0.745	1.49
31	0.828	1.27	0.799	1.34	0.747	1.47
For larger n and	$1/(1+1.645/\sqrt{2n})$		$1/(1+1.960/\sqrt{2n})$		$1/(1+2.576/\sqrt{2n})$	
	$1/(1-1.645/\sqrt{2n})$		$1/(1-1.960/\sqrt{2n})$		$1/(1-2.576/\sqrt{2n})$	

[a] Confidence limits for $\sigma = b_L s$ and $b_U s$.

lower 0.95 confidence limit of

$$\hat{\sigma}_L = 0.688(4.83) = 3.32 \; lb.$$

and upper 0.95 confidence limit of

$$\hat{\sigma}_U = 1.83(4.83) = 8.83 \; lb.$$

If we wish a one-sided confidence limit on the low side with confidence coefficient P, we estimate the lower one-sided confidence limit as

$$\hat{\sigma}_L = s\sqrt{(n-1)\big/\chi^2_{(1-P)}}$$

For a one-sided confidence limit on the high side with confidence coefficient P, we estimate the upper one-sided confidence limit as

$$\hat{\sigma}_U = s\sqrt{(n-1)\big/\chi^2_P}$$

Thus for $P = 0.95, 0.975,$ and 0.995 we use the b_L or b_U factor from Table 7 in the columns headed 0.90, 0.95, and 0.99, respectively. For example, a 0.95 upper one-sided confidence limit for σ based on a sample of 10 items for which $s = 4.83$ will be

$$\begin{aligned}\hat{\sigma}_U &= b_{U(0.90)}s \\ &= 1.64(4.83) \\ &= 7.92\end{aligned}$$

A lower 0.95 one-sided confidence limit would be

$$\begin{aligned}\hat{\sigma}_L &= b_{L(0.90)}s \\ &= 0.730(4.83) \\ &= 3.53\end{aligned}$$

SUPPLEMENT B

Presenting Plus or Minus Limits of Uncertainty for p'[4]

When there is a fraction p of a given category, for example, the fraction nonconforming, in n observations obtained under controlled conditions, 95 percent confidence limits for the population fraction p' may be found from the chart in Fig. 41 of *Biometrika Tables for Statisticians*, Vol. 1. A reproduction of this fraction is entered on the abscissa and the upper and lower 0.95 confidence limits are chart for selected sample sizes is shown in Fig. 4. To use the chart, the sample read on the vertical scale for various values of n. Approximate limits for values of n not shown on the Biometrika chart may be obtained by graphical interpolation. The *Biometrika Tables for Statisticians* also give a chart for 0.99 confidence limits.

In general, for an np and $n(1-p)$ at least 6 and preferably $0.10 \leq p \leq 0.90$, the following formulas can be applied

approximate 0.90 confidence limits

$$p \pm 1.645\sqrt{p(1-p)\big/n}$$

approximate 0.95 confidence limits

$$p \pm 1.960\sqrt{p(1-p)\big/n} \qquad (4)$$

approximate 0.99 confidence limits

$$p \pm 2.576\sqrt{p(1-p)\big/n}.$$

Example

Refer to the data of Table 2(a) of **PART 1** and Fig. 4 of **PART 1** and suppose that the lower specification limit on transverse

---

[4] The analysis is strictly valid only for an unlimited population such as presented by a manufacturing or measurement process. When the population sampled is relatively small compared with the sample size n, the reader is advised to consult a statistician.

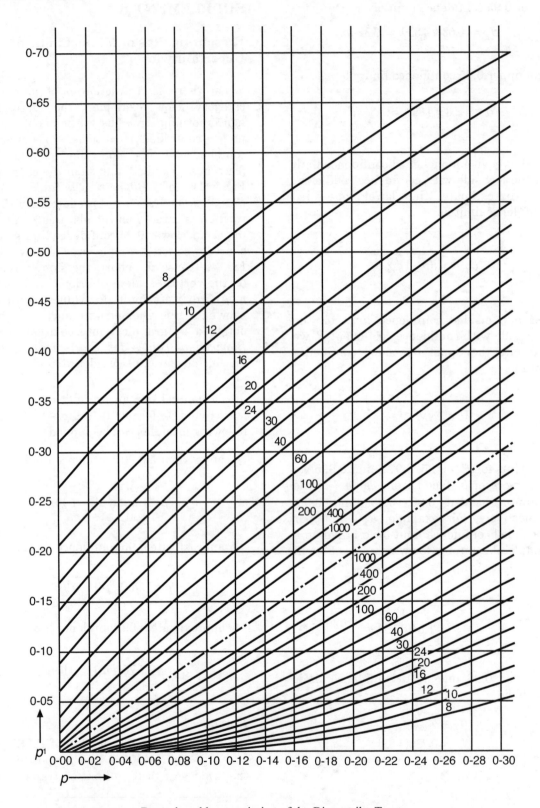

FIG. 4—Chart providing confidence limits for p' in binomial sampling, given a sample fraction. Confidence coefficient = 0.95. The numbers printed along the curves indicate the sample size n. If for a given value of the abscissa, p_A and p_B are the ordinates read from (or interpolated between) the appropriate lower and upper curves, then $\Pr\{p_A \leq p' \leq p_B\} \geq 0.95$.

strength is 675 psi and there is no upper specification limit. Then the sample percentage of bricks nonconforming (the sample "fraction nonconforming" p) is seen to be 8/270 = 0.030. Rough 0.95 confidence limits for the universe fraction nonconforming p' are read from Fig. 4 as 0.02 to 0.07. Using Eq 4, we have approximate 95 percent confidence limits as

$$0.030 \pm 1.960\sqrt{0.030(1 - 0.030/270)}$$

$$= 0.030 \pm 1.960(0.010)$$
$$= \begin{cases} 0.05 \\ 0.01 \end{cases}$$

Even though $p > 0.10$, the two results agree reasonably well.

One-sided confidence limits for a population fraction p' can be obtained directly from the Biometrika chart or Fig. 4, but the confidence coefficient will be 0.975 instead of 0.95 as in the two-sided case. For example, with $n = 200$ and the sample $p = 0.10$ the 0.975 upper one-sided confidence limit is read from Fig. 4 to be 0.15. When the Normal approximation can be used, we will have the following approximate one-sided confidence limits for p'

$P = 0.90:$ $\quad$ lower limit $= p - 1.282\sqrt{p(1-p)/n}$
$\quad\quad\quad\quad$ upper limit $= p + 1.282\sqrt{p(1-p)/n}$

$P = 0.95:$ $\quad$ lower limit $= p - 1.645\sqrt{p(1-p)/n}$
$\quad\quad\quad\quad$ upper limit $= p + 1.645\sqrt{p(1-p)/n}$

$P = 0.99:$ $\quad$ lower limit $= p - 2.326\sqrt{p(1-p)/n}$
$\quad\quad\quad\quad$ upper limit $= p + 2.326\sqrt{p(1-p)/n}$

REFERENCES

[1] Shewhart, W. A., "Probability as a Basis for Action," presented at the joint meeting of the American Mathematical Society and Section K of the A.A.A.S., 27 Dec. 1932.

[2] Shewhart, W. A., *Statistical Method from the Viewpoint of Quality Control*, W. E. Deming, Ed., The Graduate School, Department of Agriculture, Washington, D.C., 1939.

[3] Pearson, E. S., *The Application of Statistical Methods to Industrial Standardization and Quality Control*, B. S. 600-1935, British Standards Institution, London, Nov. 1935.

[4] Snedecor, G. W. and Cochran, W. G., *Statistical Methods*, 7th ed., Iowa State University Press, Ames, IA, 1980, pp. 54–56.

[5] Ott, E. R., Schilling, E. G.,. and Neubauer, D. V., *Process Quality Control*, 3rd ed., McGraw-Hill, New York, N.Y., 2000.

3

Control Chart Method of Analysis and Presentation of Data

GLOSSARY OF TERMS AND SYMBOLS USED IN PART 3

In general, the terms and symbols used in **PART 3** have the same meanings as in preceding parts of the Manual. In a few cases, which are indicated in the following glossary, a more specific meaning is attached to them for the convenience of a portion or all of **PART 3**. Mathematical definitions and derivations are given in **Supplement A**.

Glossary of Terms

assignable cause, *n.*—an identifiable factor that contributes to variation in quality and which it is feasible to detect and identify. Sometimes referred to as a *special cause*.

chance cause, *n.*—an identifiable factor that exhibits variation which is random and free from any recognizable pattern over time. Sometimes referred to as a *common cause*.

lot, *n.*—a definite quantity of some commodity produced under conditions that are considered uniform for sampling purposes.

sample, *n.*—a group of units, or portion of material, taken from a larger collection of units or quantity of material, which serves to provide information that can be used as a basis for action on the larger quantity or on the production process. May be referred to as a *subgroup* in the construction of a control chart.

subgroup, *n.*—one of a series of groups of observations obtained by subdividing a larger group of observations; alternatively, the data obtained from one of a series of samples taken from a series of lots or from sublots taken from a process. One of the essential features of the control chart method is to break up the inspection data into *rational subgroups,* that is, to classify the observed values into subgroups, *within* which variations may, for engineering reasons, be considered to be due to nonassignable chance causes only, but *between* which there may be differences due to one or more assignable causes whose presence is considered possible. May be referred to as a *sample* from the process in the construction of a control chart.

unit, *n.*—one of a number of similar articles, parts, specimens, lengths, areas, etc. of a material or product.

sublot, *n.*—an identifiable part of a lot.

Glossary of Symbols

Symbol	General	In **PART 3**, Control Charts
c	. . .	*number of nonconformities;* more specifically the number of nonconformities in a sample (subgroup)

Glossary of Symbols—*Continued*

Symbol	General	In **PART 3,** Control Charts
c_4		a factor that is a function of n and expresses the ratio between the expected value of $\bar{s}$ for a large number of samples of n observed values each and the σ of the universe sampled. (Values of $c_4 = E(\bar{s})/\sigma$ are given in Tables 6 and 16, and in Table 49 in **Supplement A,** based on a Normal distribution.)
d_2	. . .	a factor that is a function of n and expresses the ratio between expected value of $\bar{R}$ for a large number of samples of n observed values each and the σ of the universe sampled. (Values of $d_2 = E(\bar{R})/\sigma$ are given in Tables 6 and 16 and in Table 49 in **Supplement A,** based on a Normal distribution.)
k		number of subgroups or samples under consideration
MR	typically the absolute value of the difference of two successive values plotted on a control chart. It may also be the range of a group of more than two successive values.	the absolute value of the difference of two successive values plotted on a control chart.
$\overline{MR}$	*average* of n-1 moving ranges from a series of n values	*average moving range* of n-1 moving ranges from a series of n values $$\overline{MR} = \frac{\left|X_2 - X_1\right| + \cdots + \left|X_n - X_{n-1}\right|}{n-1}$$
n	*number* of observed values (observations)	subgroup or sample size, that is, the number of units or observed values in a sample or subgroup
p	*relative frequency or proportion,* the ratio of the number of occurrences to the total possible number of occurrences	*fraction nonconforming,* the ratio of the number of nonconforming units (articles, parts, specimens, etc.) to the total number of units under consideration; more specifically, the fraction nonconforming of a sample (subgroup)
np	number of occurrences	*number of nonconforming units;* more specifically, the *number of nonconforming units in* a sample of n units
R	*range* of a set of numbers, that is, the difference between the largest number and the smallest number	range of the n observed values in a subgroup (sample) (the symbol R is also used to designate the moving range in Figures 29 and 30)
s	*sample standard deviation*	standard deviation of the n observed values in a subgroup (sample) $$s = \sqrt{\frac{(X_1 - \overline{X})^2 + \cdots + (X_n - \overline{X})^2}{n-1}}$$ or expressed in a form more convenient but sometimes less accurate for computation purposes $$s = \sqrt{\frac{n(X_1^2 + \cdots + X_n^2) - (X_1 + \cdots + X_n)^2}{n(n-1)}}$$
u		*nonconformities per units,* the number of nonconformities in a sample of n units divided by n
X	observed value of a measurable characteristic; specific observed values are designated $X_1, X_2, X_3,$ etc.; also used to designate a measurable characteristic	
$\overline{X}$	*average* (arithmetic mean); the sum of the n observed values divided by n	average of the n observed values in a subgroup (sample) $$\overline{X} = \frac{X_1 + X_2 + \cdots + X_n}{n}$$

Qualified Symbols

$\sigma_{\bar{x}}, \sigma_{s}, \sigma_{R}, \sigma_{p}$; etc.	*standard deviation* of values of $\bar{X}$, s, R, p, etc.	*standard deviation* of the sampling distribution of $\bar{X}$, s, R, p, etc.
$\bar{\bar{X}}, \bar{s}, \bar{R}, \bar{p}$, etc.	*average* of a set of values of $\bar{X}$, s, R, p, etc. (the overbar notation signifies an average value)	*average* of the set of k subgroup (sample) values of $\bar{X}$ s, R, p, etc., under consideration; for samples of unequal size, an overall or weighted average
μ, σ, p′, u′, c′	*mean, standard deviation, fraction nonconforming, etc.,* of the population	
$\mu_0, p_0, u_0, c_0,$ σ_0		*standard value* of μ, σ, p′, etc., adopted for computing control limits of a control chart for the case, Control with Respect to a Given Standard (see Sections 18 to 27)
α	*alpha risk* of claiming that a hypothesis is true when it is actually true	*risk* of claiming that a process is out of statistical control when it is actually in statistical control, a.k.a., Type I error. 100(1-α)% is the *percent confidence*.
β	*beta risk* of claiming that a hypothesis is false when it is actually false	*risk* of claiming that a process is in statistical control when it is actually out of statistical control, a.k.a., Type II error. 100(1-β)% is the *power* of a test that declares the hypothesis is false when it is actually false.

GENERAL PRINCIPLES

1. Purpose

PART 3 of the Manual gives formulas, tables, and examples that are useful in applying the *control chart* method [1] of analysis and presentation of data. This method requires that the data be obtained from several samples or that the data be capable of subdivision into subgroups based on relevant engineering information. Although the principles of **PART 3** are applicable generally to many kinds of data, they will be discussed herein largely in terms of the quality of materials and manufactured products.

The control chart method provides a criterion for detecting lack of statistical control. Lack of statistical control in data indicates that observed variations in quality are greater than should be attributed to chance. Freedom from indications of lack of control is necessary for scientific evaluation of data and the determination of quality.

The control chart method lays emphasis on the *order* or grouping of the observations in a set of individual observations, sample averages, number of nonconformities, etc., with respect to time, place, source, or any other consideration that provides a basis for a classification which may be of significance in terms of known conditions under which the observations were obtained.

This concept of order is illustrated by the data in Table 1 in which the width in inches to the nearest 0.0001-in. is given for 60 specimens of Grade BB zinc which were used in ASTM atmospheric corrosion tests.

At the left of the table, the data are tabulated without regard to relevant information. At the right, they are shown arranged in ten subgroups, where each subgroup relates to the specimens from a separate milling. The information regarding origin is relevant engineering

TABLE 1. Comparison of data before and after subgrouping (width in inches of specimens of Grade BB zinc).

Before Subgrouping			After Subgrouping						
				Specimen					
0.5005	0.5005	0.4996	Subgroup (Milling)	1	2	3	4	5	6
0.5000	0.5002	0.4997							
0.5008	0.5003	0.4993							
0.5000	0.5004	0.4994	1	0.5005	0.5000	0.5008	0.5000	0.5005	0.5000
0.5005	0.5000	0.4999							
0.5000	0.5005	0.4996	2	0.4998	0.4997	0.4998	0.4994	0.4999	0.4998
0.4998	0.5008	0.4996							
0.4997	0.5007	0.4997	3	0.4995	0.4995	0.4995	0.4995	0.4995	0.4996
0.4998	0.5008	0.4995							
0.4994	0.5010	0.4995	4	0.4998	0.5005	0.5005	0.5002	0.5003	0.5004
0.4999	0.5008	0.4997							
0.4998	0.5009	0.4992	5	0.5000	0.5005	0.5008	0.5007	0.5008	0.5010
0.4995	0.5010	0.4995							
0.4995	0.5005	0.4992	6	0.5008	0.5009	0.5010	0.5005	0.5006	0.5009
0.4995	0.5006	0.4994							
0.4995	0.5009	0.4998	7	0.5000	0.5001	0.5002	0.4995	0.4996	0.4997
0.4995	0.5000	0.5000							
0.4996	0.5001	0.4990	8	0.4993	0.4994	0.4999	0.4996	0.4996	0.4997
0.4998	0.5002	0.5000							
0.5005	0.4995	0.5000	9	0.4995	0.4995	0.4997	0.4995	0.4995	0.4992
			10	0.4994	0.4998	0.5000	0.4990	0.5000	0.5000

information, which makes it possible to apply the control chart method to these data (see *Example 3*).

2. Terminology and Technical Background

Variation in quality from one unit of product to another is usually due to a very large number of causes. Those causes, for which it is possible to identify, are termed *special causes*, or *assignable causes*. Lack of control indicates one or more assignable causes are operative. The vast majority of causes of variation may be found to be inconsequential and cannot be identified. These are termed *chance causes*, or *common causes*. However, causes of large variations in quality generally admit of ready identification.

In more detail we may say that for a constant system of chance causes, the average $\overline{X}$, the standard deviations s, the value of fraction nonconforming p, or any other functions of the observations of a series of samples will exhibit statistical stability of the kind that may be expected in random samples from homogeneous material. The criterion of the quality control chart is derived from laws of chance variations for such samples, and failure to satisfy this criterion is taken as evidence of the presence of an operative assignable cause of variation.

As applied by the manufacturer to inspection data, the control chart provides a basis for *action*. Continued use of the control chart and the elimination of assignable causes as their presence is disclosed, by failures to meet its criteria, tend to reduce variability and to stabilize quality at aimed-at levels [2-8,12]. While the control chart method has been devised primarily for this purpose, it provides simple techniques and criteria that have been found useful in analyzing and interpreting other types of data as well.

3. Two Uses

The control chart method of analysis is used for the following two distinct purposes.

A. Control—No Standard Given

To discover whether observed values of $\overline{X}$, s, p, etc., for several samples of n observations each, *vary among themselves* by an amount greater than should be attributed to chance. Control charts based entirely on the data from samples are used for detecting *lack of constancy* of the cause system.

B. Control with Respect to a Given Standard

To discover whether observed values of $\overline{X}$, s, p, etc., for samples of n observations each, differ from standard values μ_0, σ_0, p_0, etc., by an amount greater than should be attributed to chance. The standard value may be an experience value based on representative prior data, or an economic value established on consideration of needs of service and cost of production, or a desired or aimed-at value designated by specification. It should be noted particularly that the standard value of σ, which is used not only for setting up control charts for s or R but also for computing control limits on control charts for $\overline{X}$, should almost invariably be an experience value based on representative *prior* data. Control charts based on such standards are used particularly in inspection to control processes and to maintain quality uniformly at the level desired.

4. Breaking up Data into Rational Subgroups

One of the essential features of the control chart method is what is referred to as breaking up the data into rationally chosen subgroups called *rational subgroups*. This means classifying the observations under consideration into subgroups or samples, *within* which the variations may be considered on engineering grounds to be due to nonassignable chance causes only, but *between* which the differences may be due to assignable causes whose presence are suspected or considered possible.

This part of the problem depends on technical knowledge and familiarity with the conditions under which the material sampled was produced and the conditions under which the data were taken. By identifying each sample with a time or a source, specific causes of trouble may be more readily traced and corrected, if advantageous and economical. Inspection and test records, giving observations in the order in which they were taken, provide directly a basis for subgrouping with respect to time. This is commonly advantageous in manufacture where it is important, from the standpoint of quality, to maintain the production cause system constant with time.

It should always be remembered that analysis will be greatly facilitated if, when planning for the collection of data in the first place, care is taken to so select the samples that the data from each sample can properly be treated as a separate rational subgroup, and that the samples are identified in such a way as to make this possible.

5. General Technique in Using Control Chart Method

The general technique (see Ref. **1**, Criterion I, Chapter XX) of the control chart method variations in quality generally admit of ready identification is as follows. Given a set of observations, to determine whether an assignable cause of variation is present:

a. Classify the total number of observations into k rational subgroups (samples) having n_1, n_2, ..., n_k observations, respectively. Make subgroups of *equal size*, if practicable. It is usually preferable to make subgroups not smaller than $n = 4$ for variables $\overline{X}$, s, or R, nor smaller than $n = 25$ for (binary) attributes. (See Sections 13, 15, 23, and 25 for further discussion of preferred sample sizes and subgroup expectancies for general attributes.)

b. For each statistic ($\overline{X}$, s, R, p, etc.) to be used, construct a control chart with control limits in the manner indicated in the subsequent sections.

c. If one or more of the observed values of $\overline{X}$, s, R, p, etc., for the k subgroups (samples) fall outside the control limits, take this fact as an indication of the presence of an assignable cause.

6. Control Limits and Criteria of Control

In both uses indicated in Section 3, the control chart consists essentially of symmetrical limits (control limits) placed above and below a central line. The central line in each case indicates the expected or average value of $\overline{X}$, s, R, p, etc. for subgroups (samples) of n observations each.

The control limits used here, referred to as *3-sigma control limits*, are placed at a distance of three standard deviations from the central line. The standard deviation is defined as the standard deviation of the sampling distribution of the statistical measure in question ($\overline{X}$, s, R, p, etc.) for subgroups (samples) of size n. Note that this standard deviation is *not* the standard deviation computed from the subgroup values (of $\overline{X}$, s, R, p, etc.) plotted on the chart, but is computed from the variations within the subgroups (see **Supplement B,** *Note 1*).

Throughout this part of the Manual such standard deviations of the sampling distributions will be designated as $\sigma_{\overline{X}}$, σ_s, σ_R, σ_p, etc., and these symbols, which consist of σ and a subscript, will be used only in this restricted sense.

For measurement data, if μ and σ were known, we would have

Control limits for:

average	(expected $\overline{X}$)$\pm 3\sigma_{\overline{x}}$
standard deviations	(expected s) $\pm 3\sigma_s$
ranges	(expected R) $\pm 3\sigma_R$

where the various expected values are derived from estimates of μ or σ. For attribute data, if p' were known, we would have control limits for values of p (expected p) $\pm 3\sigma_p$, where expected $p = p'$.

The use of 3-sigma control limits can be attributed to Walter Shewhart who based this practice upon the evaluation of numerous datasets (see Ref. **1**). Shewhart determined that based on a single point relative to 3-sigma control limits the control chart would signal assignable causes affecting the process. Use of 4-sigma control limits would not be sensitive enough, and use of 2-sigma control limits would produce too many false signals (too sensitive) based on the evaluation of a single point.

Figure 1 indicates the features of a control chart for averages. The choice of the factor 3 (a multiple of the expected standard deviation of $\overline{X}$, s, R, p, etc.) in these limits, as Shewhart suggested (Ref. **1**), is an economic choice based on experience that covers a wide range of industrial applications of the control chart, rather than on any exact value of probability (see **Supplement B**, *Note 2*). This choice has proved satisfactory for use as a criterion for *action*, that is, for looking for assignable causes of variation.

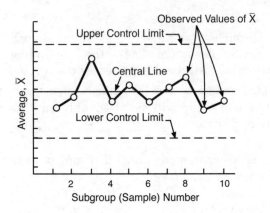

FIG. 1—Essential features of a control chart presentation; chart for averages.

This *action* is presumed to occur in the normal work setting, where the cost of too frequent false alarms would be uneconomic. Furthermore, the situation of too frequent false alarms could lead to a rejection of the control chart as a tool if such deviations on the chart are of no practical or engineering significance. In such a case, the control limits should be re-evaluated to determine if they correctly reflect the system of chance, or common, cause variation of the process. For example, a control chart on a raw material assay may have understated control limits if the data on which they were based encompassed only a single lot of the raw material. Some lot-to-lot raw material variation would be expected since nature is in control of the assay of the material as it is being mined. Of course, in some cases, some compensation by the supplier may be possible to correct problems with particle size and the chemical composition of the material in order to comply with the customer's specification.

In *exploratory* research, or in the early phases of a deliberate investigation into potential improvements, it may be worthwhile to investigate points that fall outside what some have called a set of *warning limits* (often placed two standards deviation about the centerline). The chances that any single point would fall two standard deviations from the average is roughly 1 in 20, or 5% of the time, when the process is indeed centered and in statistical control. Thus, stopping to investigate a false alarm once for every 20 plotting points on a control chart would be too excessive. Alternatively, an effective *rule of nonrandomness* would be to take action if two consecutive points were beyond the warning limits on the same side of the centerline. The *risk* of such an action would only be roughly 1 in 800! Such an occurrence would be considered an unlikely event and indicate that the process is not in control, so justifiable action would be taken to identify an assignable cause.

A control chart may be said to display a lack of control under a variety of circumstances, any of which provide some evidence of nonrandom behavior. Several of the best known nonrandom patterns can be detected by the manner in which one or more tests for nonrandomness are violated. The following list of such tests are given below:

1. Any single point beyond 3σ limits

2. Two consecutive points beyond 2σ limits on the same side of the centerline

3. Eight points in a row on one side of the centerline

4. Six points in a row that are moving away or toward the centerline with no change in direction (a.k.a., trend rule)

5. Fourteen consecutive points alternating up and down (sawtooth pattern)

6. Two out of three points beyond 2σ limits on the same side of the centerline

7. Four out of five points beyond 1σ limits on the same side of the centerline

8. Fifteen points in a row within the 1σ limits on either side of the centerline (a.k.a., stratification rule—sampling from two sources *within* a subgroup)

9. Eight consecutive points outside the 1σ limits on both sides of the centerline (a.k.a., mixture rule—sampling from two sources *between* subgroups)

There are other rules that can be applied to a control chart in order to detect nonrandomness, but those given here are the most common rules in practice.

It is also important to understand what risks are involved when implementing control charts on a process. If we state that *the process is in a state of statistical control*, and present it as a hypothesis, then we can consider what risks are operative in any process investigation. In particular, there are two types of risk that can be seen in the following table:

DECISION ABOUT THE STATE OF THE PROCESS BASED ON DATA	TRUE STATE OF THE PROCESS	
	Process is IN control	Process is OUT of control
Process is IN control	No error is made	Beta (β) risk, or Type II error
Process is OUT of control	Alpha (α) risk, or Type I error	No error is made

For a set of data analyzed by the control chart method, when may a state of control be assumed to exist? Assuming subgrouping based on time, it is usually not safe to assume that a state of control

exists unless the plotted points for at least 25 consecutive subgroups fall within 3-sigma control limits. On the other hand, the number of subgroups needed to detect a lack of statistical control at the start may be as small as 4 or 5. Such a precaution against overlooking trouble may increase the risk of a false indication of lack of control. But it is a risk almost always worth taking in order to detect trouble early.

What does this mean? If the objective of a control chart is to detect a process change, and that we want to know how to improve the process, then it would be desirable to assume a larger alpha (α) risk (smaller beta (β) risk) by using control limits smaller than 3 standard deviations from the centerline. This would imply that there would be more false signals of process change if the process were actually in control. Conversely, if the alpha risk is too small, by using control limits larger than 2 standard deviations from the centerline, then we may not be able to detect a process change when it occurs which results in a larger beta (β) risk.

Typically, in a process improvement effort, it is desirable to consider a larger alpha risk with a smaller beta risk. However, if the primary objective is to control the process with a minimum of false alarms, then it would be desirable to have a smaller alpha risk with a larger beta risk. The latter situation is preferable if the user is concerned about the occurrence of too many false alarms, and is confident that the control chart limits are the best approximation of chance cause variation.

Once statistical control of the process has been established, occurrence of one plotted point beyond 3-sigma limits in 35 consecutive subgroups or 2 points in 100 subgroups need not be considered a cause for action.

NOTE

In a number of examples in **PART 3**,

fewer than 25 points are plotted. In most of these examples, evidence of a lack of control is found. In others, it is considered only that the charts fail to show such evidence, and it is not safe to assume a state of statistical control exists.

CONTROL —NO STANDARD GIVEN

7. Introduction

Sections 7 to 17 cover the technique of analysis for control when no standard is given, as noted under *A* in Section 3. Here standard values of μ, σ, p', etc., are *not given*, hence values derived from the numerical observations are used in arriving at central lines and control limits. This is the situation that exists when the problem at hand is the analysis and presentation of a given set of experimental data. This situation is also met in the initial stages of a program using the control chart method for controlling quality during production. Available information regarding the quality level and variability resides in the data *to be analyzed* and the central lines and control limits are based on values derived from those data. For a contrasting situation, see Section 18.

8. Control Charts for Averages, $\overline{X}$, and for Standard Deviations, s—Large Samples

This section assumes that a set of observed values of a variable X can be subdivided into k rational subgroups (samples), each subgroup containing n of *more than* 25 observed values.

A. *Large Samples of Equal Size*

For samples of size n, the control chart lines are as shown in Table 2.

TABLE 2. Formulas for control chart lines.[1]

	CENTRAL LINE	CONTROL LIMITS	
For averages $\overline{X}$	$\overline{\overline{X}}$	$\overline{\overline{X}} \pm 3\dfrac{\overline{s}}{\sqrt{n-0.5}}$	(1)[a]
For standard deviations s	$\overline{s}$	$\overline{s} \pm 3\dfrac{\overline{s}}{\sqrt{2n-2.5}}$	(2)[b]

[a] Formula 1 for control limits is an approximation based on Eq 70, **Supplement A.** It may be used for n of 10 or more.

[b] Formula 2 for control limits is an approximation based on Eq 75, **Supplement A.** It may be used for n of 10 or more.

where

$\overline{\overline{X}}$ = the grand average of the observed values of X for *all* samples,

$$= (\overline{X}_1 + \overline{X}_2 + \cdots + \overline{X}_k)/k \qquad (3)$$

$\overline{s}$ = the average subgroup standard deviation,

$$= (s_1 + s_2 + \cdots + s_k)/k \qquad (4)$$

where the subscripts 1, 2, . . ., k refer to the k subgroups, respectively, all of size n. (For a discussion of this formula, see **Supplement B,** *Note 3;* also see *Example 1*).

B. *Large Samples of Unequal Size*

Use Equations 1 and 2 but compute $\overline{\overline{X}}$ and $\overline{s}$ as follows

$\overline{\overline{X}}$ = the grand average of the observed values of X for all samples

---

[1] Previous editions of this manual had used n instead of $n - 0.5$ in Equation 1, and 2 $(n - 1)$ instead of $2n - 2.5$ in Equation 2 for control limits. Both formulas are approximations, but the present ones are better for n less than 50. Also, it is important to note that the lower control limit for the standard deviation chart is the maximum of $\overline{s} - 3\sigma_s$ and 0 since negative values have no meaning. This idea also applies to the lower control limits for attribute control charts.

$$= \frac{n_1 \overline{X}_1 + n_2 \overline{X}_2 + \cdots + n_k \overline{X}_k}{n_1 + n_2 + \cdots + n_k} \qquad (5)$$

= grand total of X values divided by their total number

$\overline{s}$ = the weighted standard deviation

$$= \frac{n_1 s_1 + n_2 s_2 + \cdots + n_k s_k}{n_1 + n_2 + \cdots + n_k} \qquad (6)$$

where the subscripts 1, 2, . . . , k refer to the k subgroups, respectively. (For a discussion of this formula, see **Supplement B**, *Note 3*.) Then compute control limits for each sample size separately, using the individual sample size, n, in the formula for control limits (see *Example 2*).

When most of the samples are of approximately equal size, computing and plotting effort can be saved by the procedure given in **Supplement B**, *Note 4*.

9. Control Charts for Averages $\overline{X}$, and for Standard Deviations, s—Small Samples

This section assumes that a set of observed values of a variable X is subdivided into k rational subgroups (samples), each subgroup containing $n=25$ *or fewer* observed values.

A. Small Samples of Equal Size

For samples of size n, the control chart lines are shown in Table 3. The centerlines for these control charts are defined as the overall average of the statistics being plotted, and can be expressed as

TABLE 3. Formulas for control chart lines.[a]

	CENTRAL LINE	CONTROL LIMITS		
		FORMULA USING FACTORS IN TABLE 6	ALTERNATE FORMULA	
For averages $\overline{X}$	$\overline{\overline{X}}$	$\overline{\overline{X}} \pm A_3 \overline{s}$	$\overline{\overline{X}} \pm 3 \dfrac{\overline{s}}{\sqrt{n - 0.5}}$	(7)[a]
For standard deviations s	$\overline{s}$	$B_4\overline{s}$ and $B_3\overline{s}$	$\overline{s} \pm 3 \dfrac{\overline{s}}{\sqrt{2n - 2.5}}$	(8)[b]

[a]Alternate Formula 7 is an approximation based on Eq 70, **Supplement A**. It may be used for n of 10 or more. The values of A_3 in the tables were computed from Eqs 42 and 57 in **Supplement A**.

[b]Alternate Formula 8 is an approximation based on Eq 75, **Supplement A**. It may be used for n of 10 or more. The values of B_3 and B_4 in the tables were computed from Eqs 42, 61, and 62 in **Supplement A**.

$\overline{\overline{X}}$ = the grand average of observed values of X for *all* samples,

$$\overline{s} = \frac{s_1 + s_2 + \ldots + s_k}{k} \qquad (9)$$

and s_1, s_2, etc., refer to the observed standard deviations for the first, second, etc., samples and factors c_4, A_3, B_3, and B_4 are given in Table 6. For a discussion of Eq 9, see **Supplement B**, *Note 3*; also see *Example 3*.

B. Small Samples of Unequal Size

For *small samples* of *unequal size*, use Equations 7 and 8 (or corresponding factors) for computing control chart lines. Compute $\overline{\overline{X}}$ by Eq 5. Obtain separate derived values of $\overline{s}$ for the different sample sizes by the following working rule: Compute $\hat{\sigma}$ the overall *average* value of the observed ratio s/c_4 for the individual samples; then compute $\overline{s} = c_4\hat{\sigma}$ for each sample size n. As shown in *Example 4*, the computation can be simplified by combining in separate groups all samples having the same sample size n. Control limits may then be determined separately for each sample size. These difficulties can be avoided by planning the collection of data so that the samples are made of equal size. The factor c_4 is given in Table 6 (see *Example 4*).

10. Control Charts for Averages, $\overline{X}$, and for Ranges, R—Small Samples

This section assumes that a set of observed values of a variable X is subdivided into k rational subgroups (samples), each subgroup containing $n = 10$ or fewer observed values.

The range, R, of a sample is the difference between the largest observation and the smallest observation. When $n = 10$ or less, simplicity and economy of effort can be obtained by using control charts for $\overline{X}$ and R in place of control charts for $\overline{X}$ and s. The range is not recommended, however, for samples of more than 10 observations, since it becomes rapidly less effective than the standard deviation as a detector of assignable causes as n increases beyond this value. In some circumstances it may be found satisfactory to use the control chart for ranges for samples up to $n = 15$, as when data are plentiful or cheap. On occasion it may be desirable to use the chart for ranges for even larger samples: for this reason Table 6 gives factors for samples as large as $n = 25$.

A. Small Samples of Equal Size

For samples of size n, the control chart lines are as shown in Table 4.

TABLE 4. Formulas for control chart lines.

| | CENTRAL LINE | CONTROL LIMITS | |
		FORMULA USING FACTORS IN TABLE 6	ALTERNATE FORMULA
For Averages $\overline{X}$	$\overline{\overline{X}}$	$\overline{\overline{X}} \pm A_2\overline{R}$	$\overline{\overline{X}} \pm 3\dfrac{\overline{R}}{d_2\sqrt{n}}$ (10)
For Ranges R	$\overline{R}$	$D_4\overline{R}$ and $D_3\overline{R}$	$\overline{R} \pm 3\dfrac{d_3\overline{R}}{d_2}$ (11)

where $\overline{\overline{X}}$ is the grand average of observed values of X for *all* samples, $\overline{R}$ is the average value of range R for the k individual samples

$$(R_1 + R_2 + \ldots + R_k)/k \qquad (12)$$

and the factors d_2, A_2, D_3 and D_4 are given in Table 6, and d_3 in Table 49 (See *Example 5*).

B. Small Samples of Unequal Size

For *small samples of unequal size*, use Formulas 10 and 11 (or corresponding factors) for computing control chart lines. Compute $\overline{\overline{X}}$ by Eq 5. Obtain separate derived values of $\overline{R}$ for the different sample sizes by the following working rule: compute $\hat{\sigma}$, the overall *average* value of the observed ratio R/d_2 for the individual samples. Then compute $\overline{R} = d_2\hat{\sigma}$ for each sample size n. As shown in *Example 6*, the computation can be simplified by combining in separate groups all samples having the same sample size n. Control limits may then be determined separately for each sample size. These difficulties can be avoided by planning the collection of data so that the samples are made of equal size.

11. Summary, Control Charts for $\overline{X}$, s, and R—No Standard Given

The most useful formulas and equations from Sections 7 to 10, inclusive, are collected in Table 5 and are followed by Table 6, which gives the factors used in these and other formulas.

12. Control Charts for Attributes Data

Although in what follows the fraction p is designated *fraction nonconforming*, the methods described can be applied quite generally and p may in fact be used to represent the ratio of the number of items,

TABLE 5. Formulas for control chart lines.[a]

	CENTRAL LINE	CONTROL LIMITS
Averages using s	$\overline{\overline{X}}$	$\overline{\overline{X}} \pm A_3 \bar{s}$ ($\bar{s}$ as given by Eq 9)
Averages using R	$\overline{\overline{X}}$	$\overline{\overline{X}} \pm A_2 \bar{R}$ ($\bar{R}$ as given by Eq 12)
Standard deviations	$\bar{s}$	$B_4 \bar{s}$ and $B_3 \bar{s}$ ($\bar{s}$ as given by Eq 9)
Ranges	$\bar{R}$	$D_4 \bar{R}$ and $D_3 \bar{R}$ ($\bar{R}$ as given by Eq 12)

[a]Control—no standard given (μ, σ, not given)—small samples of equal size.

TABLE 6. Factors for computing control chart lines—no standard given.

	CHART FOR AVERAGES		CHART FOR STANDARD DEVIATIONS			CHART FOR RANGES		
	FACTORS FOR CONTROL LIMITS		FACTORS FOR CENTRAL LINE	FACTORS FOR CONTROL LIMITS		FACTORS FOR CENTRAL LINE	FACTORS FOR CONTROL LIMITS	
OBSERVATIONS IN SAMPLE, n	A_2	A_3	c_4	B_3	B_4	d_2	D_3	D_4
2	1.880	2.659	0.7979	0	3.267	1.128	0	3.267
3	1.023	1.954	0.8862	0	2.568	1.693	0	2.575
4	0.729	1.628	0.9213	0	2.266	2.059	0	2.282
5	0.577	1.427	0.9400	0	2.089	2.326	0	2.114
6	0.483	1.287	0.9515	0.030	1.970	2.534	0	2.004
7	0.419	1.182	0.9594	0.118	1.882	2.704	0.076	1.924
8	0.373	1.099	0.9650	0.185	1.815	2.847	0.136	1.864
9	0.337	1.032	0.9693	0.239	1.761	2.970	0.184	1.816
10	0.308	0.975	0.9727	0.284	1.716	3.078	0.223	1.777
11	0.285	0.927	0.9754	0.321	1.679	3.173	0.256	1.744
12	0.266	0.886	0.9776	0.354	1.646	3.258	0.283	1.717
13	0.249	0.850	0.9794	0.382	1.618	3.336	0.307	1.693
14	0.235	0.817	0.9810	0.406	1.594	3.407	0.328	1.672
15	0.223	0.789	0.9823	0.428	1.572	3.472	0.347	1.653
16	0.212	0.763	0.9835	0.448	1.552	3.532	0.363	1.637
17	0.203	0.739	0.9845	0.466	1.534	3.588	0.378	1.622
18	0.194	0.718	0.9854	0.482	1.518	3.640	0.391	1.609
19	0.187	0.698	0.9862	0.497	1.503	3.689	0.404	1.596
20	0.180	0.680	0.9869	0.510	1.490	3.735	0.415	1.585
21	0.173	0.663	0.9876	0.523	1.477	3.778	0.425	1.575
22	0.167	0.647	0.9882	0.534	1.466	3.819	0.435	1.565
23	0.162	0.633	0.9887	0.545	1.455	3.858	0.443	1.557
24	0.157	0.619	0.9892	0.555	1.445	3.895	0.452	1.548
25	0.153	0.606	0.9896	0.565	1.435	3.931	0.459	1.541
Over 25	...	a	b	c	d	...	...	...

[a] $3/\sqrt{n-0.5}$

[b] $(4n-4)/(4n-3)$

[c] $1 - 3/\sqrt{2n-2.5}$

[d] $1 + 3/\sqrt{2n-2.5}$

occurrences, etc. that possess some given attribute to the total number of items under consideration.

The fraction nonconforming, p, is particularly useful in analyzing inspection and test results that are obtained on a "go/ no-go" basis (method of attributes). In addition, it is used in analyzing results of measurements that are made on a scale and recorded (method of variables). In the latter case, p may be used to represent the fraction of the total number of measured values falling above any limit, below any limit, between any two limits, or outside any two limits.

The fraction p is used widely to represent the fraction nonconforming, that is, the ratio of the number of nonconforming units (articles, parts, specimens, etc.) to the total number of units under consideration. The fraction nonconforming is used as a measure of quality with respect to a single quality characteristic or with respect to two or more quality characteristics treated collectively. In this connection it is important to distinguish between a *nonconformity* and a *nonconforming unit*. A nonconformity is a *single* instance of a failure to meet some requirement, such as a failure to comply with a particular requirement imposed on a unit of product with respect to a single quality characteristic. For example, a unit containing departures from requirements of the drawings and specifications with respect to (1) a particular dimension, (2) finish, and (3) absence of chamfer, contains three defects. The words "nonconforming unit" define a unit (article, part, specimen, etc.) containing one or more "nonconformities" with respect to the quality characteristic under consideration.

When only a single quality characteristic is under consideration, or when only one nonconformity can occur on a unit, the number of nonconforming units in a sample will equal the number of nonconformities in that sample. However, it is suggested that under these circumstances the phrase "number of nonconforming units" be used rather than "number of nonconformities."

Control charts for attributes are usually based either on counts of occurrences or on the average of such counts. This means that a series of attribute samples may be summarized in one of these two principal forms of a control chart and, though they differ in appearance, both will produce essentially the same evidence as to the state of statistical control. Usually it is not possible to construct a second type of control chart based on the same attribute data which gives evidence different from that of the first type of chart as to the state of statistical control in the way the $\overline{X}$ and s (or $\overline{X}$ and R) control charts do for variables.

An exception may arise when, say, samples are composed of similar units in which various numbers of nonconformities may be found. If these numbers in individual units are recorded, then in principle it is possible to plot a second type of control chart reflecting variations in the number of nonuniformities from unit to unit within samples. Discussion of statistical methods for helping to judge whether this second type of chart gives different information on the state of statistical control is beyond the scope of this Manual.

In control charts for attributes, as in s and R control charts for small samples, the lower control limit is often at or near zero. A point above the upper control limit on an attribute chart may lead to a costly search for cause. It is important, therefore, especially when small counts are likely to occur, that the calculation of the upper limit accounts adequately for the magnitude of chance variation that may be expected. Ordinarily there is little to justify the use of a control chart for attributes if the occurrence of one or two nonconformities in a sample causes a point to fall above the upper control limit.

NOTE

To avoid or minimize this problem of small counts, it is best if the expected or estimated number of occurrences in a sample is four or more. An attribute control chart is least useful when the expected number of occurrences in a sample is less than one.

NOTE

The lower control limit based on the formulas given may result in a negative value that has no meaning. In such situations, the lower control limit is simply set at zero.

It is important to note that a positive non-zero lower control limit offers the opportunity for a plotted point to fall below this limit when the process quality level significantly improves. Identifying the assignable cause(s) for such points will usually lead to opportunities for process and quality improvements.

13. Control Chart for Fraction Nonconforming p

This section assumes that the total number of units tested is subdivided into k rational subgroups (samples) consisting of n_1, n_2,..., n_k units, respectively, for each of which a value of p is computed.

Ordinarily the control chart of p is most useful when the samples are large, say when n is 50 or more, and when the expected number of nonconforming units (or other occurrences of interest) per sample is four or more, that is, the

expected np is four or more. When n is less than 25 or when the expected np is less than 1, the control chart for p may not yield reliable information on the state of control.

The average fraction nonconforming $\bar{p}$ is defined as

$$\bar{p} = \frac{\text{total \# of nonconforming units in all samples}}{\text{total \# of units in all samples}}$$

= fraction nonconforming in the complete set of test results.

(13)

A. Samples of Equal Size.

For a sample of size n, the control chart lines are as follows in Table 7 (see Example 7).

TABLE 7. Formulas for control chart lines.

	CENTRAL LINE	CONTROL LIMITS	
For values of p	$\bar{p}$	$\bar{p} \pm 3\sqrt{\dfrac{\bar{p}(1-\bar{p})}{n}}$	(14)

When $\bar{p}$ is small, say less than 0.10, the factor $1 - \bar{p}$ may be replaced by unity for most practical purposes, which gives control limits for p by the simple relation

$$\bar{p} \pm 3\sqrt{\frac{\bar{p}}{n}} \qquad (14a)$$

B. Samples of Unequal Size

Proceed as for samples of equal size but compute control limits for each sample size separately.

When the data are in the form of a series of k subgroup values of p and the

TABLE 8. Formulas for control chart lines.

	CENTRAL LINE	CONTROL LIMITS
For values of np	$n\bar{p}$	$n\bar{p} \pm 3\sqrt{n\bar{p}(1-\bar{p})}$ (16)

corresponding sample sizes n, $\bar{p}$ may be computed conveniently by the relation

$$\bar{p} = \frac{n_1 p_1 + n_2 p_2 + \cdots + n_k p_k}{n_1 + n_2 + \cdots + n_k} \qquad (15)$$

where the subscripts 1, 2,..., k refer to the k subgroups. When most of the samples are of approximately equal size, computation and plotting effort can be saved by the procedure in **Supplement B, Note 4** (see *Example 8*).

NOTE

If a sample point falls above the upper control limit for p when $n\bar{p}$ is less than 4, the following check and adjustment method is recommended to reduce the incidence of misleading indications of a lack of control. If the non-integral remainder of the product of n and the upper control limit value for p is one-half or less, the indication of a lack of control stands. If that remainder exceeds one-half, add one to the product and divide the sum by n to calculate an adjusted upper control limit for p. Check for an indication of lack of control in p against this adjusted limit (see *Example 7* and *Example 8*).

14. Control Chart for Numbers of Nonconforming Units np

The control chart for np, number of conforming units in a sample of size n, is the equivalent of the control chart for p, for which it is a convenient practical substitute when all samples have the same size n. It makes direct use of the number of nonconforming units np, in a sample (np = the fraction nonconforming times the sample size.)

For samples of size n, the control chart lines are as shown in Table 8, where

$n\bar{p}$ = total number of nonconforming units in all samples/number of samples

$$(17)$$

     = the *average* number of nonconforming units in the k individual samples, and

$\bar{p}$ = the value given by Eq 13.

When $\bar{p}$ is *small*, say less than 0.10, the factor $1-\bar{p}$ may be replaced by unity for most practical purposes, which gives control limits for np by the simple relation

$$n\bar{p} \pm 3\sqrt{n\bar{p}} \qquad (18)$$

or, in other words, it can be read as the

avg. number of nonconforming units $\pm$

$$3\sqrt{\text{average number of nonconforming units}}$$

where "average number of nonconforming units" means the average number in samples of equal size (see *Example 7*).

When the sample size, n, varies from sample to sample, the control chart for p (Section 13) is recommended in preference to the control chart for np; in this case, a graphical presentation of values of np does not give an easily understood picture, since the expected values, $n\bar{p}$, (central line on the chart) vary with n, and therefore the plotted values of np become more difficult to compare. The recommendations of Section 13 as to size of n and expected np in a sample apply also to control charts for the numbers of nonconforming units.

When only a single quality characteristic is under consideration, and when only one nonconformity can occur on a unit, the word "nonconformity" can be substituted for the words "nonconforming unit" throughout the discussion of this section but this practice is not recommended.

NOTE

If a sample point falls above the upper control limit for np when $n\bar{p}$ is less than 4, the following check and adjustment procedure is to be recommended to reduce the incidence of misleading indications of a lack of control. If the non-integral remainder of the upper control limit value for np is one-half or less, the indication of a lack of control stands. If that remainder exceeds one-half, add one to the upper control limit value for np to adjust it. Check for an indication of lack of control in np against this adjusted limit (see *Example 7*).

15. Control Chart for Nonconformities per Unit u

In inspection and testing, there are circumstances where it is possible for several nonconformities to occur on a single unit (article, part, specimen, unit length, unit area, etc.) of product, and it is desired to control the number of nonconformities per unit, rather than the fraction nonconforming. For any given sample of units, the numerical value of nonconformities per unit, u, is equal to the number of nonconformities in all the units in the sample divided by the number of units in the sample.

The control chart for the nonconformities per unit in a sample u is convenient for a product composed of units for which inspection covers more than one characteristic, such as dimensions checked by gages, electrical and mechanical characteristics checked by tests, and visual nonconformities observed by the eye. Under these circumstances, several independent nonconformities may occur on one unit of product and a better measure of quality is obtained by making a count of all nonconformities observed and dividing by the number of units inspected to give a value of nonconformities per unit, rather than merely counting the number of nonconforming units to give a value of fraction nonconforming. This is particularly the case for complex assemblies where the occurrence of two or more nonconformities on a unit may be relatively frequent. However, only *independent* nonconformities are counted. Thus, if two nonconformities occur on one unit of product and the second is caused by the first, only the first is counted.

The control chart for nonconformities per unit (more especially the chart for number of nonconformities, see Section 16) is a particularly convenient one to use when the number of possible nonconformities on a unit is indeterminate, as for physical defects (finish or surface irregularities, flaws, pinholes, etc.) on such products as textiles, wire, sheet materials, etc., which are not continuous or extensive. Here, the opportunity for nonconformities may be numerous though the chances of nonconformities occurring at any one spot may be small.

Table 9. Formulas for control chart lines.

	CENTRAL LINE	CONTROL LIMITS	
For values of u	$\bar{u}$	$\bar{u} \pm 3\sqrt{\dfrac{\bar{u}}{n}}$	(20)

This section assumes that the total number of units tested is subdivided into k rational subgroups (samples) consisting of $n_1, n_2, \ldots, n_k$ units, respectively, for each of which a value of u is computed.

The control chart for u is most useful when the expected nu is 4 or more. When the expected nu is less than 1, the control chart for u may not yield reliable information on the state of control.

The average nonconformities per unit, $\bar{u}$, is defined as

$$\bar{u} = \frac{\text{total \# nonconformities in all samples}}{\text{total \# units in all samples}}$$
$$= \text{nonconformities per unit in the} \qquad (19)$$
$$\text{complete set of test results}$$

The simplified relations shown for control limits for nonconformities per unit assume that for each of the characteristics under consideration the ratio of the expected number of nonconformities to the possible number of nonconformities is small, say less than 0.10, an assumption that is commonly satisfied in quality control work. For an explanation of the nature of the distribution involved, see **Supplement B**, *Note 5*.

A. Samples of Equal Size

For samples of size n (n = number of units), the control chart lines are as shown in Table 9.

For samples of equal size, a chart for the number of nonconformities, c, is recommended, see Section 16. In the special case where each sample consists of only one unit, that is, $n = 1$, then the chart for u (nonconformities per unit) is identical with that chart for c (number of nonconformities) and may be handled in accordance with Section 16. In this case the chart may be referred to either as a chart for nonconformities per unit or as a chart for number of nonconformities, but the latter designation is recommended (see *Example 9*).

B. Samples of Unequal Size

Proceed as for samples of equal size but compute the control limits for each sample size separately.

When the data are in the form of a series of subgroup values of u and the corresponding sample sizes, $\bar{u}$ may be computed by the relation

$$\bar{u} = \frac{n_1 u_1 + n_2 u_2 + \cdots + n_k u_k}{n_1 + n_2 + \cdots + n_k} \qquad (21)$$

where as before, the subscripts 1, 2, ..., k refer to the k subgroups.

Note that n_1, n_2, etc., need not be whole numbers. For example, if u represents nonconformities per 1000 ft of wire, samples of 4000 ft, 5280 ft, etc., then the corresponding values will be 4.0, 5.28, etc., units of 1000 ft.

When most of the samples are of approximately equal size, computing and plotting effort can be saved by the procedure in **Supplement B**, *Note 4* (see *Example 10*).

NOTE

If a sample point falls above the upper limit for u where $n\bar{u}$ is less than 4, the following check and adjustment procedure is recommended to reduce the incidence of misleading indications of a lack of control. If the non-integral remainder of the product of n and the upper control limit value for u is one half or less, the indication of a lack of control stands. If that remainder exceeds one-half, add one to the product and divide the sum by n to calculate an adjusted upper control limit for u. Check for an indication of lack of control in u against this adjusted limit (see *Examples 9* and *10*).

16. Control Chart for Number of Nonconformities, *c*

The control chart for c, the number of nonconformities in a sample, is the equivalent of the control chart for u, for which it is a convenient practical substitute when all samples have the *same size n* (number of units).

A. Samples of Equal Size

For samples of equal size, if the average number of nonconformities per sample is $\bar{c}$, the control chart lines are as shown in Table 10.

TABLE 10. Formulas for control chart lines.

	CENTRAL LINE	CONTROL LIMITS	
For values of c	$\bar{c}$	$\bar{c} \pm 3\sqrt{\bar{c}}$	(22)

where

$$\bar{c} = \frac{\text{total number of nonconformities in all samples}}{\text{number of samples}}$$
$$= \text{average number of nonconformities per sample.}$$
(23)

The use of c is especially convenient when there is no natural unit of product, as for nonconformities over a surface or along a length, and where the problem is to determine uniformity of quality in equal lengths, areas, etc., of product (see *Example 9* and *Example 11*).

B. Samples of Unequal Size

For samples of unequal size, first compute the average nonconformities per unit $\bar{u}$, by Eq 19; then compute the control limits for each sample size separately as shown in Table 11.

TABLE 11. Formulas for control chart lines.

	CENTRAL LINE	CONTROL LIMITS	
For values of c	$n\bar{u}$	$n\bar{u} \pm 3\sqrt{n\bar{u}}$	(24)

The control chart for u is recommended as preferable to the control chart for c when the sample size varies from sample to sample for reasons stated in discussing the control charts for p and np. The recommendations of Section 15 as to expected $c = n\bar{u}$ also applies to control charts for numbers of nonconformities.

TABLE 12. Formulas for control chart lines.

CONTROL—NO STANDARD GIVEN—ATTRIBUTES DATA			
	CENTRAL LINE	CONTROL LIMITS	APPROXIMATION
Fraction nonconforming p	$\bar{p}$	$\bar{p} \pm 3 \sqrt{\dfrac{\bar{p}(1-\bar{p})}{n}}$	$\bar{p} \pm 3 \sqrt{\dfrac{\bar{p}}{n}}$
Number of nonconforming units np	$n\bar{p}$	$n\bar{p} \pm 3 \sqrt{n\bar{p}(1-\bar{p})}$	$n\bar{p} \pm 3 \sqrt{n\bar{p}}$
Nonconformities per unit, u	$\bar{u}$	$\bar{u} \pm 3 \sqrt{\dfrac{\bar{u}}{n}}$	...
Number of nonconformities c samples of equal size	$\bar{c}$	$\bar{c} \pm 3 \sqrt{\bar{c}}$	...
samples of unequal size	$n\bar{u}$	$n\bar{u} \pm 3 \sqrt{n\bar{u}}$	...

NOTE

If a sample point falls above the upper control limit for c when $n\bar{u}$ is less than 4, the following check and adjustment procedure is to be recommended to reduce the incidence of misleading indications of a lack of control. If the non-integral remainder of the upper control limit for c is one-half or less, the indication of a lack of control stands. If that remainder exceeds one-half, add one to the upper control limit value for c to adjust it. Check for an indication of lack of control in c against this adjusted limit (see *Examples 9 and 11*).

17. Summary, Control Charts for *p*, *np*, *u*, and *c*—No Standard Given

The formulas of Sections 13 to 16, inclusive, are collected as shown in Table 12 for convenient reference.

CONTROL WITH RESPECT TO A GIVEN STANDARD

18. Introduction

Sections 18 to 27 cover the technique of analysis for control with respect to a given standard, as noted under (B) in Section 3. Here, standard values of μ, σ, $p,'$ etc., are *given*, and are those corresponding to a given standard distribution. These standard values, designated as μ_0, σ_0, p_0 etc., are used in calculating both central lines and control limits. (When only μ_0 is given and no prior data are available for establishing a value of σ_0, analyze data from the first production period as in Sections 7 to 10, but use μ_0 for the central line.)

Such standard values are usually based on a control chart analysis of previous data (for the details, see **Supplement B**, *Note 6*), but may be given on the basis described in Section 3*B*. Note that these standard values are set up *before* the detailed analysis of the data at hand is undertaken and frequently before the data to be analyzed are collected. In addition to the standard values, only the information regarding sample size or sizes is required in order to compute central lines and control limits.

For example, the values to be used as central lines on the control charts are

for averages,	μ_0
for standard deviations,	$c_4\sigma_0$
for ranges,	$d_2\sigma_0$
for values of p,	p_0
etc.,	

where factors c_4 and d_2, which depend only on the sample size, n, are given in Table 16, and defined in **Supplement A**.

Note that control with respect to a given standard may be a more exacting requirement than control with no standard given, described in Sections 7 to 17. The data must exhibit not only control but control at a standard level and with no more than standard variability.

Extending control limits obtained from a set of existing data into the future and using these limits as a basis for purposive control of quality during production, is equivalent to adopting, as standard, the values obtained from the existing data. Standard values so obtained may be tentative and subject to revision as more experience is accumulated (for details, see **Supplement B**, *Note 6*).

NOTE

Two situations that are not covered specifically within this section should be mentioned.

1. In some cases a standard value of μ is given as noted above, but no standard value is given for σ. Here σ is estimated from the analysis of the data at hand and the problem is essentially one of

controlling $\overline{X}$ at the standard level μ_0 that has been given.

2. In other cases, interest centers on controlling the conformance to specified minimum and maximum limits within which material is considered acceptable, sometimes established without regard to the actual variation experienced in production. Such limits may prove unrealistic when data are accumulated and an estimate of the standard deviation, say σ^* of the process is obtained therefrom. If the natural spread of the process (a band having a width of $6\sigma^*$), is wider than the spread between the specified limits, it is necessary either to adjust the specified limits or to operate within a band narrower than the process capability. Conversely, if the spread of the process is narrower than the spread between the specified limits, the process will deliver a more uniform product than required. Note that in the latter event when only maximum and minimum limits are specified, the process can be operated at a level above or below the indicated mid-value without risking the production of significant amounts of unacceptable material.

19. Control Charts for Averages $\overline{X}$ and for Standard Deviation, s

For samples of size n, the control chart lines are as shown in Table 13.

For samples of n greater than 25, replace c_4 by $(4n-4)/(4n-3)$.

See *Examples 12* and *13*; also see **Supplement B**, *Note 9*.

TABLE 13. Formulas for control chart lines.[2]

	CENTRAL LINE	CONTROL LIMITS		
		FORMULA USING FACTORS IN TABLE 16	ALTERNATE FORMULA	
For averages $\overline{X}$	μ_0	$\mu_0 \pm A\sigma_0$	$\mu_0 \pm 3\dfrac{\sigma_0}{\sqrt{n}}$	(25)
For standard deviations s	$c_4\sigma_0$	$B_6\sigma_0$ and $B_4\sigma_0$	$c_4\sigma_0 \pm \dfrac{\sigma_0}{\sqrt{2n-1.5}}$	(26)[a]

[a] Alternate Formula 26 is an approximation based on Eq 74, **Supplement A**. It may be used for n of 10 or more. The values of B_5 and B_6 given in the tables are computed from Eqs 42, 59, and 60 in **Supplement A**.

For samples of n = 25 or less, use Table 16 for factors A, B_5, and B_6. Factors c_4, A, B_5 and B_6 are defined in **Supplement A**. See *Examples 14* and *15*.

20. Control Chart for Ranges R

The range, R, of a sample is the difference between the largest observation and the smallest observation.

For samples of size n, the control chart lines are as shown in Table 14.

TABLE 14. Formulas for control chart lines.

	CENTRAL LINE	CONTROL LIMITS		
		FORMULA USING FACTORS IN TABLE 16	ALTERNATE FORMULA	
For range R	$d_2\sigma_0$	$D_2\sigma_0$ and $D_1\sigma_0$	$d_2\sigma_0 \pm d_3\sigma_0$	(27)

Use Table 16 for factors d_2, D_1, and D_2.

Factors d_2, d_3, D_1, and D_2 are defined in **Supplement A**.

---

[2]Previous editions of this manual had $2(n-1)$ instead of $2n-1.5$ in Alternate Formula 26. Both formulas are approximations, but the present one is better for n less than 50

For comments on the use of the control chart for ranges, see Section 10 (also see *Example 16*).

21. Summary, Control Charts for $\overline{X}$, s, and $\overline{R}$ —Standard Given

The most useful formulas from Sections 19 and 20 are summarized as shown in Table 15 and are followed by Table 16 which gives the factors used in these and other formulas.

22. Control Charts for Attributes Data

The definitions of terms and the discussions in Sections 12 to 16, inclusive, on the use of the fraction nonconforming, p, number of nonconforming units, np, nonconformities per unit, u, and number of nonconformities, c, as measures of quality are equally applicable to the sections which follow and will not be repeated here. It will suffice to discuss the central lines and control limits when standards are given.

23. Control Chart for Fraction Nonconforming, p

Ordinarily, the control chart for p is most useful when samples are large, say, when n is 50 or more and when the expected number of nonconforming units (or other occurrences of interest) per sample is four or more, that is, the expected values of np is four or more. When n is less than 25 or the expected np is less than 1, the control chart for p may not yield reliable information on the state of control even with respect to a given standard.

TABLE 15. Formulas for control chart lines

CONTROL WITH RESPECT TO A GIVEN STANDARD (μ_0, σ_0 GIVEN)		
	CENTRAL LINE	CONTROL LIMITS
Average $\overline{X}$	μ_0	$\mu_0 \pm A\sigma_0$
Standard Deviations s	$c_4\sigma_0$	$B_6\sigma_0$ and $B_5\sigma_0$
Range R	$d_2\sigma_0$	$D_2\sigma_0$ and $D_1\sigma_0$

TABLE 16. Factors for computing control chart lines—standard given.

OBSERVATIONS IN SAMPLE, n	CHART FOR AVERAGES — FACTORS FOR CONTROL LIMITS — A	CHART FOR STANDARD DEVIATIONS — FACTOR FOR CENTRAL LINE — c_4	FACTORS FOR CONTROL LIMITS — B_5	B_6	CHART FOR RANGES — FACTOR FOR CENTRAL LINE — d_2	FACTORS FOR CONTROL LIMITS — D_1	D_2
2	2.121	0.7979	0	2.606	1.128	0	3.686
3	1.732	0.8862	0	2.276	1.693	0	4.358
4	1.500	0.9213	0	2.088	2.059	0	4.698
5	1.342	0.9400	0	1.964	2.326	0	4.918
6	1.225	0.9515	0.029	1.874	2.534	0	5.079
7	1.134	0.9594	0.113	1.806	2.704	0.205	5.204
8	1.061	0.9650	0.179	1.751	2.847	0.388	5.307
9	1.000	0.9693	0.232	1.707	2.970	0.547	5.393
10	0.949	0.9727	0.276	1.669	3.078	0.686	5.469
11	0.905	0.9754	0.313	1.637	3.173	0.811	5.535
12	0.866	0.9776	0.346	1.610	3.258	0.923	5.594
13	0.832	0.9794	0.374	1.585	3.336	1.025	5.647
14	0.802	0.9810	0.399	1.563	3.407	1.118	5.696
15	0.775	0.9823	0.421	1.544	3.472	1.203	5.740
16	0.750	0.9835	0.440	1.526	3.532	1.282	5.782
17	0.728	0.9845	0.458	1.511	3.588	1.356	5.820
18	0.707	0.9854	0.475	1.496	3.640	1.424	5.856
19	0.688	0.9862	0.490	1.483	3.689	1.489	5.889
20	0.671	0.9869	0.504	1.470	3.735	1.549	5.921
21	0.655	0.9876	0.516	1.459	3.778	1.606	5.951
22	0.640	0.9882	0.528	1.448	3.819	1.660	5.979
23	0.626	0.9887	0.539	1.438	3.858	1.711	6.006
24	0.612	0.9892	0.549	1.429	3.895	1.759	6.032
25	0.600	0.9896	0.559	1.420	3.931	1.805	6.056
Over 25	$3/\sqrt{n}$	a	b	c	...	...	...

$^a (4n-4)/(4n-3)$

$^b (4n-4)/(4n-3) - 3/\sqrt{2n-2.5}$

$^c (4n-4)/(4n-3) + 3/\sqrt{2n-2.5}$

See **Supplement B**, *Note 9*, on replacing first term in footnotes b and c by unity.

For samples of size n, where p_0 is the standard value of p, the control chart lines are as shown in Table 17 (see *Example 17*).

TABLE 17. Formulas for control chart lines.

	CENTRAL LINE	CONTROL LIMITS	
For values of p	p_0	$p_0 \pm 3\sqrt{\dfrac{p_0(1-p_0)}{n}}$	(28)

When p_0 *is small*, say less than 0.10, the factor $1 - p_0$ may be replaced by unity for most practical purposes, which gives the simple relation for computing the control limits for p as

$$p = p_0 \pm 3\sqrt{\frac{p_0}{n}} \qquad (28a)$$

For samples of unequal size, proceed as for samples of equal size but compute control limits for each sample size separately (see *Example 18*).

When detailed inspection records are maintained, the control chart for p may be broken down into a number of component charts with advantage (see *Example 19*). See the **NOTE** at the end of Section 13 for possible adjustment of the upper control limit when np_0 is less than 4. (Substitute np_0 for $n\bar{p}$.) See *Examples 17, 18,* and *19* for applications.

24. Control Chart for Number of Nonconforming Units, *np*

The control chart for np, number of nonconforming units in a sample, is the equivalent of the control chart for fraction nonconforming, p, for which it is a convenient practical substitute, particularly when all samples have the *same size, n*. It makes direct use of the number of nonconforming units, np, in a sample (np = the product of the sample size and the fraction nonconforming). See *Example 17*.

For samples of size n, where p_0 is the standard value of p, the control chart lines are as shown in Table 18.

TABLE 18. Formulas for control chart lines.

	CENTRAL LINE	CONTROL LIMITS	
For values of np	np_0	$np_0 \pm 3\sqrt{np_0(1-p_0)}$	(29)

When p_0 *is small*, say less than 0.10, the factor $1 - p_0$ may be replaced by unity for most practical purposes, which gives the simple relation for computing the control limits for np as

$$np_0 \pm 3\sqrt{np_0} \qquad (30)$$

As noted in Section 14, the control chart for p is recommended as preferable to the control chart for np when the sample size varies from sample to sample. The recommendations of Section 23 as to size of n and the expected np in a sample also apply to control charts for the number of nonconforming units.

When only a single quality characteristic is under consideration, and when only one nonconformity can occur on a unit, the word "nonconformity" can be substituted for the words "nonconforming unit" throughout the discussion of this article, but this practice is not recommended. See the **NOTE** at the end of Section 14 for possible adjustment of the upper control limit when np_0 is less than 4. (Substitute np_0 for $n\bar{p}$.) See *Examples 17* and *18*.

25. Control Chart for Nonconformities per Unit, *u*

For samples of size n (n = number of units), where u_0 is the standard value of u, the control chart lines are as shown in Table 19.

TABLE 19. Formulas for control chart lines.

	CENTRAL LINE	CONTROL LIMITS	
For values of u	u_0	$u_0 \pm 3\sqrt{\dfrac{u_0}{n}}$	(31)

See *Examples 20* and *21*.

As noted in Section 15, the relations given here assume that for each of the characteristics under consideration, the ratio of the expected to the possible number of nonconformities is small, say less than 0.10.

If u represents "nonconformities per 1000 ft of wire," a "unit" is 1000 ft of wire. Then if a series of samples of 4000 ft are involved, u_0 represents the standard or expected number of nonconformities per 1000 ft, and $n = 4$. Note that n need not be a whole number, for if samples comprise 5280 ft of wire each, $n = 5.28$, that is, 5.28 units of 1000 ft (see *Example 11*).

Where each sample consists of only one unit, that is $n = 1$, then the chart for u (nonconformities per unit) is identical with the chart for c (number of nonconformities) and may be handled in accordance with Section 26. In this case the chart may be referred to either as a chart for nonconformities per unit or as a chart for number of nonconformities, but the latter practice is recommended.

Ordinarily, the control chart for u is most useful when the expected nu is 4 or more. When the expected nu is less than 1, the control chart for u may not yield reliable information on the state of control even with respect to a given standard.

See the **NOTE** at the end of Section 15 for possible adjustment of the upper control limit when nu_0 is less than 4. (Substitute nu_0 for $n\overline{u}$.) See *Examples 20* and *21*.

26. Control Chart for Number of Nonconformities, c

The control chart for c, number of nonconformities in a sample, is the equivalent of the control chart for nonconformities per unit for which it is a convenient practical substitute when all samples have the *same size, n* (number of units). Here c is the *number of nonconformities in a sample*.

If the standard value is expressed in terms of number of nonconformities per sample of some given size, that is, expressed merely as c_0, and the samples are all of the same given size (same number of product units, same area of opportunity for defects, same sample length of wire, etc.), then the control chart lines are as shown in Table 20.

TABLE 20. Formulas for control chart lines (c_0 given).

	CENTRAL LINE	CONTROL LIMITS	
For number of nonconformities, c	c_0	$c_0 \pm 3\sqrt{c_0}$	(32)

Use of c_0 is especially convenient when there is no natural unit of product, as for nonconformities over a surface or along a length, and where the problem of interest is to compare uniformity of quality in samples of the same size, no matter how constituted (see *Example 21*).

When the sample size, n, (number of units) varies from sample to sample, and the standard value is expressed in terms of nonconformities per unit, the control chart lines are as shown in Table 21.

TABLE 21. Formulas for control chart lines (u_0 given).

	CENTRAL LINE	CONTROL LIMITS	
For values of c	nu_0	$nu_0 \pm 3\sqrt{nu_0}$	(33)

TABLE 22. Formulas for control chart lines.

	CONTROL WITH RESPECT TO A GIVEN STANDARD (p_0, np_0, u_0 OR c_0 GIVEN)		
	CENTRAL LINE	CONTROL LIMITS	APPROXIMATION
Fraction nonconforming, p	p_0	$p_0 \pm 3\sqrt{\dfrac{p_0(1-p_0)}{n}}$	$p_0 \pm 3\sqrt{\dfrac{p_0}{n}}$
Number of nonconforming units, np	np_0	$np_0 \pm 3\sqrt{np_0(1-p_0)}$	$np_0 \pm 3\sqrt{np_0}$
Nonconformities per unit, u	u_0	$u_0 \pm 3\sqrt{\dfrac{u_0}{n}}$	
Number of nonconformities, c Samples of equal size c_0 given)	c_0	$c_0 \pm 3\sqrt{c_0}$	
Samples of unequal size (u_0 given)	nu_0	$nu_0 \pm 3\sqrt{nu_0}$	

Under these circumstances the control chart for u (Section 25) is recommended in preference to the control chart for c, for reasons stated in Section 14 in the discussion of control charts for p and for np. The recommendations of Section 25 as to the expected $c = nu$ also applies to control charts for nonconformities.

See the **NOTE** at the end of Section 16 for possible adjustment of the upper control limit when nu_0 is less than 4. (Substitute $c_0 = nu_0$ for $n\bar{u}$). See *Example 21.*

27. Summary, Control Charts for *p, np, u,* and *c*—Standard Given

The formulas of Sections 22 to 26, inclusive, are collected as shown in Table 22 for convenient reference.

CONTROL CHARTS FOR INDIVIDUALS

28. Introduction

Sections 28 to 30[3] deal with control charts for individuals, in which individual observations are plotted one by one. This type of control chart has been found useful more particularly in process control when only one observation is obtained per lot or batch of material or at periodic intervals from a process. This situation often arises when: (*a*) sampling or testing is destructive, (*b*) costly chemical analyses or physical tests are involved, and (*c*) the material sampled at any one time (such as a batch) is normally quite homogeneous, as for a well-mixed fluid or aggregate.

The purpose of such control charts is to discover whether the individual observed values differ from the expected

---

[3] To be used with caution if the distribution of individual values is markedly asymmetrical.

value by an amount greater than should be attributed to chance.

When there is some definite rational basis for grouping the batches or observations into rational subgroups, as, for example, four successive batches in a single shift, the method shown in Section 29 may be followed. In this case, the control chart for individuals is merely an adjunct to the more usual charts but will react more quickly to a sharp change in the process than the $\overline{X}$ chart. This may be important when a single batch represents a considerable sum of money.

When there is no definite basis for grouping data, the control limits may be based on the variation between batches, as described in Section 30. A measure of this variation is obtained from moving ranges of two observations each (the absolute value of successive differences between individual observations that are arranged in chronological order).

A control chart for moving ranges may be prepared as a companion to the chart for individuals, if desired, using the formulas of Section 30. It should be noted that adjacent moving ranges are correlated, as they have one observation in common.

The methods of Sections 29 and 30 may be applied appropriately in some cases where more than one observation is obtained per lot or batch, as for example with very homogeneous batches of materials, for instance, chemical solutions, batches of thoroughly mixed bulk materials, etc., for which repeated measurements on a single batch show the within-batch variation (variation of quality within a batch and errors of measurement) to be very small as compared with between-batch variation. In such cases, the *average* of the several observations for a batch may be *treated as*

an individual observation. However, this procedure should be used with great caution; the restrictive conditions just cited should be carefully noted.

The control limits given are three sigma control limits in all cases.

29. Control Chart for Individuals, X—Using Rational Subgroups

Here the control chart for individuals is commonly used as an adjunct to the more usual $\overline{X}$ and s, or $\overline{X}$ and R, control charts. This can be useful, for example, when it is important to react immediately to a single point that may be out of statistical control, when the ability to localize the source of an individual point that has gone out of control is important, or when a rational subgroup consisting of more than two points is either impractical or nonsensical. Proceed exactly as in Sections 9 to 11 (control—no standard given) or Sections 19 to 21 (control—standard given), whichever is applicable, and prepare control charts for $\overline{X}$ and s, or for $\overline{X}$ and R. In addition, prepare a control chart for individuals having the same central line as the $\overline{X}$ chart but compute the control limits as shown in Table 23.

Table 26 gives values of E_2 and E_3 for samples of $n = 10$ or less. Values that are more complete are given in Table 50, **Supplement A** for n through 25 (see *Examples 22* and *23*).

30. Control Chart for Individuals, X—Using Moving Ranges

A. No Standard Given

Here the control chart lines are computed from the observed data. In this section the symbol, *MR*, is used to signify the moving

TABLE 23. Formulas for control chart lines.

CHART FOR INDIVIDUALS—ASSOCIATED WITH CHART FOR
s OR R HAVING SAMPLE SIZE n

NATURE OF DATA	CENTRAL LINE	CONTROL LIMITS FORMULA USING FACTORS IN TABLE 26	ALTERNATE FORMULA	
NO STANDARD GIVEN				
Samples of equal size				
based on s	$\overline{X}$	$\overline{X} \pm E_3\overline{s}$	$\overline{X} \pm 3\overline{s}/c_4$	(34)
based on R	$\overline{X}$	$\overline{X} \pm E_2\overline{R}$	$\overline{X} \pm 3\overline{R}/d_2$	(35)
Samples of unequal size: σ computed from observed values of s per Section 9 or from observed values of R per Section 10(b)	$\overline{X}$		$\overline{X} \pm 3\hat{\sigma}$	(36)[a]
STANDARD GIVEN				
Samples of equal or unequal size	μ_0		$\mu_0 \pm 3\sigma_0$	(37)

[a]See Example 4 for determination of $\hat{\sigma}$ based on values of s and Example 6 for determination of $\hat{\sigma}$ based on values of R

TABLE 24. Formulas for control chart lines.

CHART FOR INDIVIDUALS—USING MOVING RANGES—
NO STANDARD GIVEN

	CENTRAL LINE	CONTROL LIMITS	
For individuals	$\overline{X}$	$\overline{X} \pm E_2\overline{MR} = \overline{X} \pm 2.66\overline{MR}$	(38)
For moving ranges of two observations	$\overline{R}$	$D_4\overline{MR} = 3.27\overline{MR}$	(39)
		$D_3\overline{MR} = 0$	

range. The control chart lines are as shown in Table 24 where

$\overline{X} = $ the average of the individual observations,

$\overline{MR} = $ the mean moving range, (see **Supplement B**, *Note 7* for more general discussion) the average of the absolute values of successive differences between pairs of the individual observations, and

$n = 2$ for determining E_2, D_3 and D_4.

See *Example 24*.

B. Standard Given

When μ_0 and σ_0 are given, the control chart lines are as shown in Table 25.

See *Example 25*.

EXAMPLES

31. Illustrative Examples — Control, No Standard Given

Examples 1 to 11, inclusive, illustrate the use of the control chart method of

TABLE 25. Formulas for control chart lines.

CHART FOR INDIVIDUALS—STANDARD GIVEN		
	CENTRAL LINE	CONTROL LIMITS
For individuals	μ_0	$\mu_0 \pm 3\sigma_0$ (40)
For moving ranges of two observations	$d_2\sigma_0$	$D_2\sigma_0 = 3.69\sigma_0$ (41) $D_1\sigma_0 = 0$

TABLE 26. Factors for computing control limits.

CHART FOR INDIVIDUALS—ASSOCIATED WITH CHART FOR s OR R HAVING SAMPLE SIZE n									
OBSERVATIONS IN SAMPLES OF EQUAL SIZE (FROM WHICH $\overline{s}$ OR $\overline{R}$ HAS BEEN DETERMINED)	2	3	4	5	6	7	8	9	10
Factors for control limits									
E_3	3.760	3.385	3.256	3.192	3.153	3.127	3.109	3.095	3.084
E_2	2.659	1.772	1.457	1.290	1.184	1.109	1.054	1.010	0.975

TABLE 27. Operating characteristic, daily control data

Sample	Sample Size n	Average $\overline{x}$	Standard Deviation s
1	50	35.1	5.35
2	50	34.6	4.73
3	50	33.2	3.73
4	50	34.8	4.55
5	50	33.4	4.00
6	50	33.9	4.30
7	50	34.4	4.98
8	50	33.0	5.30
9	50	32.8	3.29
10	50	34.8	3.77
Total	500	340.0	44.00
Average	50	34.0	4.40

analyzing data for control, when no standard is given (see Sections 7 to 17).

Example 1: Control Charts for $\overline{X}$ and s, Large Samples of Equal Size (Section 8A)

A manufacturer wished to determine if his product exhibited a state of control. In this case, the central lines and control limits were based solely on the data. Table 27 gives observed values of $\overline{X}$ and s for daily samples of $n = 50$ observations each for ten consecutive days. Figure 2 gives the control charts for $\overline{X}$ and s.

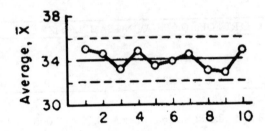

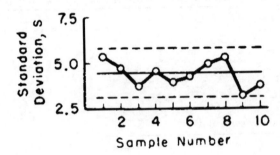

FIG. 2—Control charts for $\overline{X}$ and s. Large samples of equal size, $n = 50$; no standard given.

Central Lines

For $\overline{X} : \overline{\overline{X}} = 34.0$

For $s : \overline{s} = 4.40$

Control Limits

$n = 50$

For $\overline{X} : \overline{\overline{X}} \pm 3 \dfrac{\overline{s}}{\sqrt{n-0.5}} = 34.0 \pm 1.9,$

32.1 and 35.9

For $s : \overline{s} \pm 3 \dfrac{\overline{s}}{\sqrt{2n-2.5}} = 4.40 \pm 1.34,$

3.06 and 5.74

Results—The charts give no evidence of lack of control. Compare with *Example 12*, in which the same data are used to test product for control at a specified level.

Example 2: Control Charts for $\overline{X}$ and s, Large Samples of Unequal Size (Section 8B)

To determine whether there existed any assignable causes of variation in quality for an important operating characteristic of a given product, the inspection results given in Table 28 were obtained from ten shipments whose samples were unequal in size; hence, control limits were computed separately for each sample size.

TABLE 28. Operating characteristic, mechanical part.

Shipment	Sample Size n	Average $\overline{x}$	Standard Deviation s
1	50	55.7	4.35
2	50	54.6	4.03
3	100	52.6	2.43
4	25	55.0	3.56
5	25	53.4	3.10
6	50	55.2	3.30
7	100	53.3	4.18
8	50	52.3	4.30
9	50	53.7	2.09
10	50	54.3	2.67
Total	550	$\Sigma n\,\overline{X} = 29590.0$	$\Sigma ns = 1864.50$
Weighted average	55	53.8	3.39

Figure 3 gives the control charts for $\overline{X}$ and s.

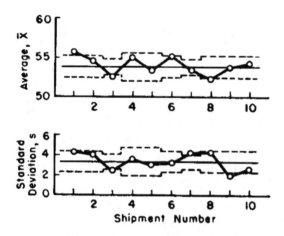

FIG. 3—Control charts for $\overline{X}$ and s. Large samples of unequal size, n = 25, 50, 100; no standard given.

Central Lines
For $\overline{X} : \overline{\overline{X}} = 53.8$
For $s : \overline{s} = 3.39$

Control Limits

For $\overline{X} : \overline{\overline{X}} \pm 3 \dfrac{\overline{s}}{\sqrt{n-0.5}} = 53.8 \pm \dfrac{10.17}{\sqrt{n-0.5}}$

$n = 25 : 51.7$ and 55.9
$n = 50 : 52.4$ and 55.2
$n = 100 : 52.8$ and 54.8

For $s : \overline{s} \pm 3 \dfrac{\overline{s}}{\sqrt{2n-2.5}} = 3.39 \pm \dfrac{10.17}{\sqrt{2n-2.5}}$

$n = 25 : 1.91$ and 4.87
$n = 50 : 2.36$ and 4.42
$n = 100 : 2.67$ and 4.11

Results—Lack of control is indicated with respect to both $\overline{X}$ and s. Corrective action is needed to reduce the variability between shipments.

Example 3: Control Charts for $\overline{X}$ and s, Small Samples of Equal Size (Section 9A)

Table 29 gives the width in inches to the nearest 0.0001-in. measured prior to exposure for ten sets of corrosion specimens of Grade BB zinc. These two groups of five sets each were selected for illustrative purposes from a large number

TABLE 29. Width in inches, specimens of Grade BB zinc.

Set	Measured Values						Average, X	Standard Deviation, s	Range, R
	X_1	X_2	X_3	X_4	X_5	X_6			
	Group 1								
1	0.5005	0.5000	0.5008	0.5000	0.5005	0.5000	0.50030	0.00035	0.0008
2	0.4998	0.4997	0.4998	0.4994	0.4999	0.4998	0.49973	0.00018	0.0005
3	0.4995	0.4995	0.4995	0.4995	0.4995	0.4996	0.49952	0.00004	0.0001
4	0.4998	0.5005	0.5005	0.5002	0.5003	0.5004	0.50028	0.00026	0.0007
5	0.5000	0.5005	0.5008	0.5007	0.5008	0.5010	0.50063	0.00035	0.0010
	Group 2								
6	0.5008	0.5009	0.5010	0.5005	0.5006	0.5009	0.50078	0.00019	0.0005
7	0.5000	0.5001	0.5002	0.4995	0.4996	0.4997	0.49985	0.00029	0.0007
8	0.4993	0.4994	0.4999	0.4996	0.4996	0.4997	0.49958	0.00021	0.0006
9	0.4995	0.4995	0.4997	0.4992	0.4995	0.4992	0.49943	0.00020	0.0005
10	0.4994	0.4998	0.5000	0.4990	0.5000	0.5000	0.49970	0.00041	0.0010
Average							0.49998	0.00025	0.00064

of sets of specimens consisting of six specimens each used in atmosphere exposure tests sponsored by ASTM. In each of the two groups, the five sets correspond to five different millings that were employed in the preparation of the specimens. Figure 4 shows control charts for $\overline{X}$ and s.

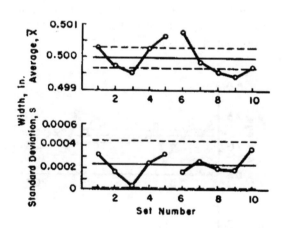

FIG. 4—Control chart for $\overline{X}$ and s. Small samples of equal size, n = 6 no standard given.

Results—The chart for averages indicates the presence of assignable causes of variation in width, $\overline{X}$, from set to set, that is, from milling to milling. The pattern of points for averages indicates a systematic pattern of width values for the five millings, a factor that required recognition in the analysis of the corrosion test results.

Central Lines

For $\overline{X}$: $\overline{\overline{X}} = 0.49998$
For s: $\overline{s} = 0.00025$

Control Limits
$n = 6$
For $\overline{X}$: $\overline{\overline{X}} \pm A_3\overline{s} =$

$0.49998 \pm (1.287)(0.00025)$

0.49966 and 0.50030

For s: $B_4\overline{s} = (1.970)(0.00025) = 0.00049$
$B_3\overline{s} = (0.030)(0.00025) = 0.00001$

Example 4: Control Charts for $\overline{X}$ and s, Small Samples of Unequal Size (Section 9B)

Table 30 gives interlaboratory calibration check data on 21 horizontal tension testing machines. The data represent tests on No. 16 wire. The procedure is similar to that given in *Example 3*, but indicates a suggested method of computation when the samples are not equal in size. Figure 5 gives control charts for $\overline{X}$ and s.

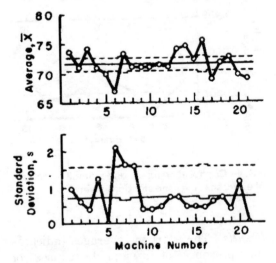

FIG. 5—Control chart for $\overline{X}$ and s. Small samples of unequal size, $n = 4$ no standard given.

$$\hat{\sigma} = \frac{1}{21}\left(\frac{2.41}{0.9213} + \frac{15.34}{0.9400}\right) = 0.902$$

Central Lines

For $\overline{X}$: $\overline{\overline{X}} = 71.65$
For s: $n = 4$: $\overline{s} = c_4\hat{\sigma} = (0.9213)(0.902)$
$= 0.831$

$n = 5$: $\overline{s} = c_4\hat{\sigma} = (0.9400)(0.902)$
$= 0.848$

Control Limits

For $\overline{X}$: $n = 4$: $\overline{\overline{X}} \pm A_3\overline{s} =$
$71.65 \pm (1.628)(0.831)$,
73.0, and 70.3

$n = 5$: $\overline{\overline{X}} \pm A_3\overline{s} =$
$71.65 \pm (1.427)(0.848)$,
72.9, and 70.4

For s: $n = 4$: $B_4\overline{s} = (2.266)(0.831) = 1.88$
$B_3\overline{s} = (0)(0.831) = 0$
$n = 5$: $B_4\overline{s} = (2.089)(0.848) = 1.77$
$B_3\overline{s} = (0)(0.848) = 0$

Results—The calibration levels of machines were not controlled at a common level; the averages of six machines are above and the averages of five machines are below the control limits. Likewise, there is an indication that the variability *within* machines is not in statistical control, since three machines, Numbers 6, 7, and 8, have standard deviations outside the control limits.

Example 5: Control Charts for $\overline{X}$ and R, Small Samples of Equal Size (Section 10A)

Same data as in *Example 3*, Table 29. Use is made of control charts for averages and ranges rather than for averages and standard deviations. Figure 6 shows control charts for $\overline{X}$ and R.

TABLE 30. Interlaboratory calibration, horizontal tension testing machines.

Machine	Number of Tests	Test Value					Average $\overline{X}$	Standard Deviation s		Range R	
		1	2	3	4	5		$n=4$	$n=5$	$n=4$	$n=5$
1	5	73	73	73	75	75	73.8	...	1.10	...	2
2	5	70	71	71	71	72	71.0	...	0.71	...	2
3	5	74	74	74	74	75	74.2	...	0.45	...	1
4	5	70	70	70	72	73	71.0	...	1.41	...	3
5	5	70	70	70	70	70	70.0	...	0	...	0
6	5	65	65	66	69	70	67.0	...	2.35	...	5
7	4	72	72	74	76	...	73.5	1.91	...	4	...
8	5	69	70	71	73	73	71.2	...	1.79	...	4
9	5	71	71	71	71	72	71.2	...	0.45	...	1
10	5	71	71	71	71	72	71.2	...	0.45	...	1
11	5	71	71	72	72	72	71.6	...	0.55	...	1
12	5	70	71	71	72	72	71.2	...	0.55	...	2
13	5	73	74	74	75	75	74.2	...	0.84	...	2
14	5	74	74	75	75	75	74.6	...	0.55	...	1
15	5	72	72	72	73	73	72.4	...	0.55	...	1
16	4	75	75	75	76	...	75.3	0.50	...	1	...
17	5	68	69	69	69	70	69.0	...	0.71	...	2
18	5	71	71	72	72	73	71.8	...	0.84	...	2
19	5	72	73	73	73	73	72.8	...	0.45	...	1
20	5	68	69	70	71	71	69.8	...	1.30	...	3
21	5	69	69	69	69	69	69.0	...	0	...	0
Total	103	weighted average $\overline{X}=71.65$						2.41	15.34	5	34

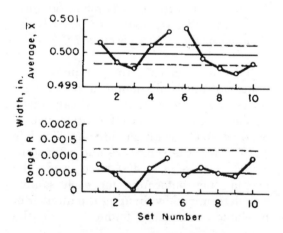

FIG. 6—Control charts for $\overline{X}$ and R. Small Samples of equal size, $n = 6$; no standard given.

Results—The results are practically identical in all respects with those obtained by using averages and standard deviations, Fig. 4, *Example 3*.

Central Lines

For $\overline{X}$: $\overline{\overline{X}} = 0.49998$

For R: $\overline{R} = 0.00064$

Control Limits

$n = 6$

For $\overline{X} : \overline{\overline{X}} \pm A_2 \overline{R}$

$= 0.49998 \pm (0.483)(0.00064)$

$= 0.50029$ and 0.49967

For $R : D_4 \overline{R} = (2.004)(0.00064) = 0.00128$

$D_3 \overline{R} = (0)(0.00064) = 0$

Example 6: Control Charts for $\overline{X}$ and R, Small Samples of Unequal Size (Section 10B)

Same data as in *Example 4*, Table 8. In the analysis and control charts, the range is used instead of the standard deviation. The procedure is similar to that given in *Example 5*, but indicates a suggested method of computation when samples are not equal in size. Figure 7 gives control charts for $\overline{X}$ and R.

$\hat{\sigma}$ is determined from the tabulated ranges given in *Example 4*, using a similar procedure to that given in *Example 4* for standard deviations where samples are not equal in size, that is

$$\hat{\sigma} = \frac{1}{21}\left(\frac{5}{2.059} + \frac{34}{2.326}\right) = 0.812$$

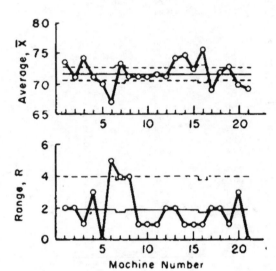

FIG. 7—Control charts for $\overline{X}$ and R. Small samples of unequal size, $n = 4,\ 5$; no standard given.

Results—The results are practically identical in all respects with those obtained by using averages and standard deviations, Fig. 5, *Example 4*.

Central Lines

For $\overline{X}$: $\overline{\overline{X}} = 71.65$

For R: $n = 4$: $\overline{R} = d_2\hat{\sigma} =$

$$(2.059)(0.812) = 1.67$$

$$n = 5:\ \overline{R} = d_2\hat{\sigma} =$$
$$(2.326)(0.812) = 1.89$$

Control Limits

For $\overline{X}$: $n = 4$: $\overline{\overline{X}} \pm A_2\overline{R} =$
$$71.65 \pm (0.729)(1.67)$$
$$70.4 \text{ and } 72.9$$

$$n = 5:\ \overline{\overline{X}} \pm A_2\overline{R} =$$
$$71.65 \pm (0.577)(1.89)$$
$$70.6 \text{ and } 72.7$$

For R : $n = 4$: $D_4\overline{R} = (2.282)(1.67) = 3.8$
$$D_3\overline{R} = (0)(1.67) = 0$$

$$n = 5:\ D_4\overline{R} = (2.114)(1.89) = 4.0$$
$$D_3\overline{R} = (0)(1.89) = 0$$

Example 7: Control Charts for p, Samples of Equal Size (Section 13A) and np, Samples of Equal Size (Section 14)

Table 31 gives the number of nonconforming units found in inspecting a series of 15 consecutive lots of galvanized washers for finish nonconformities such as exposed steel, rough galvanizing. The lots were nearly the same size and a constant sample size of $n = 400$ were used. The fraction nonconforming for each sample was determined by dividing the number of nonconforming units found, np, by the

TABLE 31. Finish defects, galvanized washers.

LOT	SAMPLE SIZE n	NUMBER OF NONCONFORMING UNITS np	FRACTION NONCONFORMING p	LOT	SAMPLE SIZE n	NUMBER OF NONCONFORMING UNITS np	FRACTION NONCONFORMING p
No. 1	400	1	0.0025	No. 9	400	8	0.0200
No. 2	400	3	0.0075	No. 10	400	5	0.0125
No. 3	400	0	0				
No. 4	400	7	0.0175	No. 11	400	2	0.0050
No. 5	400	2	0.0050	No. 12	400	0	0
				No. 13	400	1	0.0025
No. 6	400	0	0	No. 14	400	0	0
No. 7	400	1	0.0025	No. 15	400	3	0.0075
No. 8	400	0	0	Total	6000	33	0.0825

sample size, n, and is listed in the table. Figure 8 gives the control chart for p, and Fig. 9 gives the control chart for np.

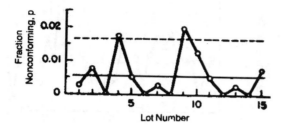

Fig. 8—Control chart for p. Samples of equal size, n=400; no standard given.

Note that these two charts are identical except for the vertical scale.

(*a*) Control chart for *p*

Central Line

$$\bar{p} = \frac{33}{6000} = 0.0055$$

$$\bar{p} = \frac{0.0825}{15} = 0.0055$$

Control Limits
$$n = 400$$

$$\bar{p} \pm 3\sqrt{\frac{\bar{p}(1-\bar{p})}{n}} =$$

$$0.0055 \pm 3\sqrt{\frac{0.0055(0.9945)}{400}} =$$

$$0.0055 \pm 0.0111 =$$

$$0 \text{ and } 0.0166$$

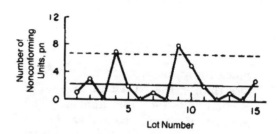

Fig. 9—Control chart for np. Samples of equal size, n=400; no standard given.

Results—Lack of control is indicated; points for lots numbers 4 and 9 are outside the control limits.

(*b*) Control chart for *np*

TABLE 32. Surface defects, galvanized hardware.

LOT	SAMPLE SIZE, n	NUMBER OF NONCONFORMING UNITS np	FRACTION NONCONFORMING p	LOT	SAMPLE SIZE n	NUMBER OF NONCONFORMING UNITS np	FRACTION NONCONFORMING p
No. 1	580	9	0.0155	No.16	330	4	0.0121
No. 2	550	7	0.0127	No.17	330	2	0.0061
No. 3	580	3	0.0052	No.18	640	4	0.0063
No. 4	640	9	0.0141	No.19	580	7	0.0121
No. 5	880	13	0.0148	No.20	550	9	0.0164
No. 6	880	14	0.0159	No.21	510	7	0.0137
No. 7	640	14	0.0219	No.22	640	12	0.0188
No. 8	550	10	0.0182	No.23	300	8	0.0267
No. 9	580	12	0.0207	No.24	330	5	0.0152
No. 10	880	14	0.0159	No.25	880	18	0.0205
No. 11	800	6	0.0075	No.26	880	7	0.0080
No. 12	800	12	0.0150	No.27	800	8	0.0100
No. 13	580	7	0.0121	No.28	580	8	0.0138
No. 14	580	11	0.0190	No.29	880	15	0.0170
No. 15	550	5	0.0091	No.30	880	3	0.0034
				No.31	330	5	0.0152
				Total	19 510	268	

Central Line
$$n = 400$$
$$n\overline{p} = \frac{33}{15} = 2.2$$

Control Limits
$$n = 400$$
$$n\overline{p} \pm 3\sqrt{n\overline{p}} = 2.2 \pm 4.4$$
$$0 \text{ and } 6.6$$

NOTE

Since the value of np is 2.2, which is less than 4, the **NOTE** at the end of Section 13 (or 14) applies. The product of n and the upper control limit value for p is $400 \times 0.0166 = 6.64$. The non-integral remainder, 0.64, is greater than one-half, and so the adjusted upper control limit value for p is $(6.64 + 1)/400 = 0.0191$. Therefore, only the point for Lot 9 is outside limits. For np, by the **NOTE** of Section 14, the adjusted upper control limit value is 7.6 with the same conclusion.

Example 8: Control Chart for p, Samples of Unequal Size (Section 13B)

Table 32 gives inspection results for surface defects on 31 lots of a certain type of galvanized hardware. The lot sizes varied considerably and corresponding variations in sample sizes were used. Figure 10 gives the control chart for fraction nonconforming p. In practice, results are commonly expressed in "percent nonconforming," using scale values of 100 times p.

Central Line
$$\overline{p} = \frac{268}{19\,510} = 0.01374$$

Control Limits
$$\overline{p} \pm 3\sqrt{\frac{\overline{p}(1-\overline{p})}{n}}$$

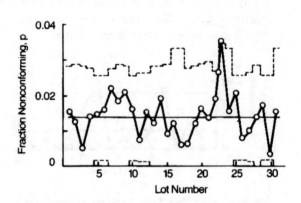

FIG. 10—Control chart for p. Samples of unequal size, n = 200 to 880; no standard given.

For $n = 300$
$$0.01374 \pm 3\sqrt{\frac{0.01374(0.98626)}{300}} =$$
$$0.01374 \pm 3(0.006720) =$$
$$0.01374 \pm 0.02016$$
$$0 \text{ and } 0.03390$$

For $n = 880$
$$0.01374 \pm 3\sqrt{\frac{0.01374(0.98626)}{880}} =$$
$$0.01374 \pm 3(0.003924) =$$
$$0.01374 \pm 0.01177$$
$$0.00197 \text{ and } 0.02551$$

Results—A state of control may be assumed to exist since 25 consecutive subgroups fall within 3-sigma control limits. There are no points outside limits, so that the **NOTE** of Section 13 does not apply.

Example 9: Control Charts for u, Samples of Equal Size (Section 15A) and c, Samples of Equal Size (Section 16A)

Table 33 gives inspection results in terms of nonconformities observed in the inspection of 25 consecutive lots of burlap bags. Since the number of bags in each lot differed slightly, a constant sample size, n

TABLE 33. Number of nonconformities in consecutive samples of ten units each—burlap bags.

SAMPLE	TOTAL NONCONFORMITIES IN SAMPLE c	NONCONFORMITIES PER UNIT u	SAMPLE	TOTAL NONCONFORMITIES IN SAMPLE c	NONCONFORMITIES PER UNIT u
1	17	1.7	13	8	0.8
2	14	1.4	14	11	1.1
3	6	0.6	15	18	1.8
4	23	2.3	16	13	1.3
5	5	0.5	17	22	2.2
6	7	0.7	18	6	0.6
7	10	1.0	19	23	2.3
8	19	1.9	20	22	2.2
9	29	2.9	21	9	0.9
10	18	1.8	22	15	1.5
11	25	2.5	23	20	2.0
12	5	0.5	24	6	0.6
			25	24	2.4
			Total	375	37.5

= 10 was used. All nonconformities were counted although two or more nonconformities of the same or different kinds occurred on the same bag. The nonconformities per unit value for each sample was determined by dividing the number of nonconformities found by the sample size and is listed in the table. Figure 11 gives the control chart for u, and Fig. 12 gives the control chart for c. Note that these two charts are identical except for the vertical scale.

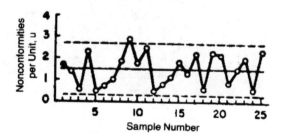

FIG. 11—Control chart for u. Samples of equal size, $n = 10$; no standard given.

(a) u
Central Line

$$\bar{u} = \frac{37.5}{25} = 1.5$$

Control Limits
$n = 10$

$$\bar{u} \pm 3\sqrt{\frac{\bar{u}}{n}} =$$

$$1.50 \pm 3\sqrt{0.150} =$$

$$1.50 \pm 1.16$$

0.34 and 2.66

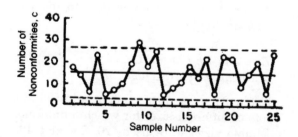

FIG. 12—Control chart for c. Samples of equal size, $n = 10$; no standard given.

(b) c
Central Line

$$\bar{c} = \frac{375}{25} = 15.0$$

TABLE 34. Number of nonconformities in samples from 20 successive lots of Type A machines.

LOT	SAMPLE SIZE n	TOTAL NON-CONFORMITIES SAMPLE c	NON-CONFORMITIES PER UNIT u	LOT	SAMPLE SIZE n	TOTAL NON-CONFORMITIES SAMPLE c	NON-CONFORMITIES PER UNIT u
No. 1	20	72	3.60	No. 11	25	47	1.88
No. 2	20	38	1.90	No. 12	25	55	2.20
No. 3	40	76	1.90	No. 13	25	49	1.96
No. 4	25	35	1.40	No. 14	25	62	2.48
No. 5	25	62	2.48	No. 15	25	71	2.84
No. 6	25	81	3.24	No. 16	20	47	2.35
No. 7	40	97	2.42	No. 17	20	41	2.05
No. 8	40	78	1.95	No. 18	20	52	2.60
No. 9	40	103	2.58	No. 19	40	128	3.20
No. 10	40	56	1.40	No. 20	40	84	2.10
				Total	580	1334	

Control Limits

$n = 10$

$$\bar{c} \pm 3\sqrt{\bar{c}} =$$

$$15.0 \pm 3\sqrt{15} =$$

$$15.0 \pm 11.6$$

$$3.4 \text{ and } 26.6$$

Results—Presence of assignable causes of variation is indicated by Sample 9. Since the value of nu is 15 (greater than 4), the **NOTE** at the end of Section 15 (or 16) does not apply.

Example 10: Control Chart for u, Samples of Unequal Size (Section 15B)

Table 34 gives inspection results for 20 lots of different sizes for which 3 different sample sizes were used, 20, 25, and 40. The observed nonconformities in this inspection cover all of the specified characteristics of a complex machine (Type A), including a large number of dimensional, operational, as well as physical and finish requirements. Because of the large number of tests and measurements required as well as possible occurrences of any minor observed irregularities, the expectancy of nonconformities per unit is high, although the majority of such nonconformities are of minor seriousness. The nonconformities per unit value for each sample, number of nonconformities in sample divided by number of units in sample, was determined and these values are listed in the last column of the table. Figure 13 gives the control chart for u with control limits corresponding to the three different sample sizes.

Central Line

$$\bar{u} = \frac{1334}{580} = 2.30$$

Control Limits

$n = 20$

$$\bar{u} \pm 3\sqrt{\frac{\bar{u}}{n}} = 2.30 \pm 1.02,$$

$$1.28 \text{ and } 3.32$$

$n = 25$

$$\bar{u} \pm 3\sqrt{\frac{\bar{u}}{n}} = 2.30 \pm 0.91,$$

$$1.39 \text{ and } 3.21$$

$n = 40$

$$\bar{u} \pm 3\sqrt{\frac{\bar{u}}{n}} = 2.30 \pm 0.72,$$

$$1.58 \text{ and } 3.02$$

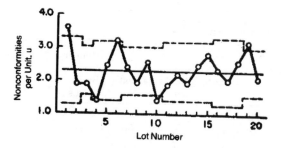

FIG. 13—Control chart for u. Samples of unequal size, n = 20, 25, 40; no standard given.

Results—Lack of control of quality is indicated; plotted points for lot numbers 1, 6, and 19 are above the upper control limit and the point for lot number 10 is below the lower control limit. Of the lots with points above the upper control limit, lot no. 1 has the smallest value of nu (46), which exceeds 4, so that the **NOTE** at the end of Section 15 does not apply.

Example 11: Control Charts for c, Samples of Equal Size (Section 16A)

Table 35 gives the results of continuous testing of a certain type of rubber-covered wire at specified test voltage. This test causes breakdowns at weak spots in the insulation, which are cut out before shipment of wire in short coil lengths. The original data obtained consisted of records of the number of breakdowns in successive lengths of 1000 ft each. There may be 0, 1, 2, 3, ..., etc. breakdowns per length, depending on the number of weak spots in the insulation. Such data might also have been tabulated as number of breakdowns in successive lengths of 100 ft each, 500 ft each, etc. Here there is no natural unit of product (such as 1 in., 1 ft, 10 ft, 100 ft, etc.), in respect to the quality characteristic "breakdown" since failures may occur at any point. Since the original data were given in terms of 1000-ft lengths, a control chart might have been

maintained for "number of breakdowns in successive lengths of 1000 ft each." So many points were obtained during a short period of production by using the 1000-ft length as a unit and the expectancy in terms of number of breakdowns per length was so small that longer unit lengths were tried. Table 35 gives (*a*) the "number of breakdowns in successive lengths of 5000 ft each," and (b) the "number of breakdowns in successive lengths of 10 000 ft each." Figure 14 shows the control chart for *c* where the unit selected is 5000ft, and Fig. 15 shows the control chart for *c* where the unit selected is 10 000 ft. The standard unit length finally adopted for control purposes was 10 000 ft for "breakdown."

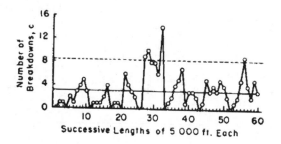

FIG. 14—Control chart for c. Samples of equal size, n =1 standard length of 5000 ft; no standard given.

(a) Lengths of 5 000 ft Each

Central Line
$$\bar{c} = \frac{187}{60} = 3.12$$

Control Limits
$$\bar{c} \pm 3\sqrt{\bar{c}} =$$
$$6.23 \pm 3\sqrt{6.23}$$
0 and 13.72

TABLE 35. Number of breakdowns in successive lengths of 5 000 ft each and 10 000 ft each for rubber-covered wire.

Length No.	Number of Breakdowns	Length No.	Number of Breakdowns	Length No.	Number of Breakdowns	Length No.	Number of Breakdowns	Length No.	Number of Breakdowns
(a) Lengths of 5 000 ft Each									
1	0	13	1	25	0	37	5	49	5
2	1	14	1	26	0	38	7	50	4
3	1	15	2	27	9	39	1	51	2
4	0	16	4	28	10	40	3	52	0
5	2	17	0	29	8	41	3	53	1
6	1	18	1	30	8	42	2	54	2
7	3	19	1	31	6	43	0	55	5
8	4	20	0	32	14	44	1	56	9
9	5	21	6	33	0	45	5	57	4
10	3	22	4	34	1	46	3	58	2
11	0	23	3	35	2	47	4	59	5
12	1	24	2	36	4	48	3	60	3
Total								60	187
(b) Lengths of 10 000 ft Each									
1	1	7	2	13	0	19	12	25	9
2	1	8	6	14	19	20	4	26	2
3	3	9	1	15	16	21	5	27	3
4	7	10	1	16	20	22	1	28	14
5	8	11	10	17	1	23	8	29	6
6	1	12	5	18	6	24	7	30	8
Total								30	187

(a) *Results*—Presence of assignable causes of variation is indicated by length numbers 27, 28, 32, and 56 falling above the upper control limit. Since the value of $\bar{c} = n\bar{u}$ is 3.12 (less than 4), the **NOTE** at the end of Section 16 does apply. The non-integral remainder of the upper control limit value is 0.42. The upper control limit stands, as do the indications of lack of control.

(b) Lengths of 10 000 ft Each

Central Line
$$\bar{c} = \frac{187}{30} = 6.23$$

Control Limits
$$\bar{c} \pm 3\sqrt{\bar{c}} =$$
$$6.23 \pm 3\sqrt{6.23}$$
$$0 \text{ and } 13.72$$

(b) *Results*—Presence of assignable causes of variation is indicated by length numbers 14, 15, 16, and 28 falling above the upper control limit. Since the value of $\bar{c}$ is 6.23 (greater than 4), the **NOTE** at the end of Section 16 does not apply.

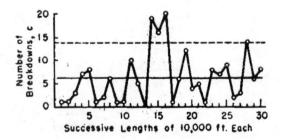

FIG. 15—Control chart for c. Samples of equal size, n = 1 *standard* length of 10 000 ft; no standard given.

32. Illustrative Examples—Control With Respect to a Given Standard

Examples 12 to 21, inclusive, illustrate the use of the control chart method of analyzing data for control with respect to a given standard (see Sections 18 to 27).

Example 12: Control Charts for $\overline{X}$ and s, Large Samples of Equal Size (Section 19)

A manufacturer attempted to maintain an aimed-at distribution of quality for a certain operating characteristic. The objective standard distribution which served as a target was defined by standard values: $\mu_0 = 35.00$ lb., and $\sigma_0 = 4.20$ lb. Table 36 gives observed values of $\overline{X}$ and s for daily samples of $n = 50$ observations each for ten consecutive days. These data are the same as used in *Example 1* and presented as Table 27. Figure 16 gives control charts for $\overline{X}$ and s.

Central Lines
For $\overline{X} : \mu_0 = 35.00$
For $s : \sigma_0 = 4.20$

Control Limits
$n = 50$
For $\overline{X}$: $\mu_0 \pm 3\dfrac{\sigma_0}{\sqrt{n}} = 35.00 \pm 1.8,$

33.2 and 36.8

For s $\left(\dfrac{4n-4}{4n-3}\right)\sigma_0 \pm 3\dfrac{\sigma_0}{\sqrt{2n-15}} = 4.18 \pm 1.27,$

2.91 and 5.45

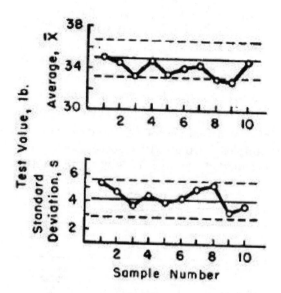

FIG. 16—Control charts for $\overline{X}$ and s. Large samples of equal size, $n = 50$; μ_0, σ_0 given.

Results—Lack of control at standard level is indicated on the eighth and ninth days. Compare with *Example 1* in which the same data were analyzed for control without specifying a standard level of quality.

TABLE 36. Operating characteristic, daily control data.

Sample	Sample Size, n	Average, $\overline{X}$	Standard Deviation, s
1	50	35.1	5.35
2	50	34.6	4.73
3	50	33.2	3.73
4	50	34.8	4.55
5	50	33.4	4.00
6	50	33.9	4.30
7	50	34.4	4.98
8	50	33.0	5.30
9	50	32.8	3.29
10	50	34.8	3.77

Example 13: Control Charts for $\overline{X}$ and s, Large Samples of Unequal Size (Section 19)

For a product, it was desired to control a certain critical dimension, the diameter, with respect to day to day variation. Daily sample sizes of 30, 50, or 75 were selected and measured, the number taken depending on the quantity produced per day. The desired level was $\mu_0 = 0.20000$ in. with $\sigma_0 = 0.00300$ in. Table 37 gives observed values of $\overline{X}$ and s for the samples from 10 successive days' production. Figure 17 gives the control charts for $\overline{X}$ and s.

TABLE 37. Diameter in inches, control data.

Sample	Sample Size n	Average $\overline{X}$	Standard Deviation s
1	30	0.20133	0.00330
2	50	0.19886	0.00292
3	50	0.20037	0.00326
4	30	0.19965	0.00358
5	75	0.19923	0.00313
6	75	0.19934	0.00306
7	75	0.19984	0.00299
8	50	0.19974	0.00335
9	50	0.20095	0.00221
10	30	0.19937	0.00397

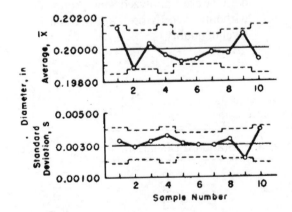

FIG. 17—Control charts for $\overline{X}$ and s. Large samples of unequal size, n = 30, 50, 70; μ_0, σ_0 given.

Central Lines

For $\overline{X}$: $\mu_0 = 0.20000$

For s: $\sigma_0 = 0.00300$

Control Limits

For $\overline{X}$: $\mu_0 \pm 3\dfrac{\sigma_0}{\sqrt{n}}$

$n = 30$

$0.2000 \pm 3\dfrac{0.00300}{\sqrt{30}} =$

0.20000 ± 0.00164

0.19836 and 0.20164

$n = 50$

0.19873 and 0.20127

$n = 75$

0.19896 and 0.20104

For s : $c_4\sigma_0 \pm 3\dfrac{\sigma_0}{\sqrt{2n-1.5}}$

$n = 30$

$\left(\dfrac{116}{117}\right) 0.00300 \pm 3\dfrac{0.00300}{\sqrt{58.5}} =$

0.00297 ± 0.00118

0.00180 and 0.00415

$n = 50$

0.00389 and 0.00208

$n = 75$

0.00225 and 0.00373

Results—The charts give no evidence of significant deviations from standard values.

Example 14: Control Chart for $\overline{X}$ and s, Small Samples of Equal Size (Section 19)

Same product and characteristic as in *Example 13*, but in this case it is desired to control the diameter of this product with respect to sample variations during each day, since samples of 10 were taken at definite intervals each day. The desired level is $\mu_0 = 0.20000$ in. with $\sigma_0 = 0.00300$ in. Table 38 gives observed values of $\overline{X}$ and s for 10 samples of 10 each taken during a single day. Figure 18 gives the control charts for $\overline{X}$ and s.

TABLE 38. Control data for one day's product.

Sample	Sample Size n	Average $\bar{x}$	Standard Deviation s
1	10	0.19838	0.00350
2	10	0.20126	0.00304
3	10	0.19868	0.00333
4	10	0.20071	0.00337
5	10	0.20050	0.00159
6	10	0.20137	0.00104
7	10	0.19883	0.00299
8	10	0.20218	0.00327
9	10	0.19868	0.00431
10	10	0.19968	0.00356

Central lines

For $\bar{X}$: $\mu_0 = 0.20000$

$$n = 10$$

For s: $c_4\sigma_0 = (0.9727)(0.00300) = 0.00292$

Control Limits

$$n = 10$$

For $\bar{X}$: $\mu_0 \pm A\sigma_0 =$

$$0.20000 \pm (0.949)(0.00300),$$

0.19715 and 0.20285

For s: $B_6\sigma_0 = (1.669)(0.00300) = 0.00501$

$B_5\sigma_0 = (0.276)(0.00300) = 0.00083$

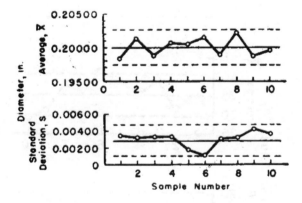

FIG.18—Control charts for $\bar{X}$ and *s*. Small samples of equal size, $n = 10$; μ_0, σ_0 given.

Results—No lack of control indicated.

Example 15: Control Chart for $\bar{X}$ and s, *Small Samples of Unequal Size (Section 19)*

A manufacturer wished to control the resistance of a certain product after it had been operating for 100 h, where $\mu_0 = 150\ \Omega$ and $\sigma_0 = 7.5\ \Omega$ from each of 15 consecutive lots, he selected a random sample of 5 units and subjected them to the operating test for 100 h. Due to mechanical failures, some of the units in the sample failed before the completion of 100 h of operation. Table 39 gives the averages and standard deviations for the 15 samples together with their sample sizes. Figure 19 gives the control charts for $\bar{X}$ and *s*.

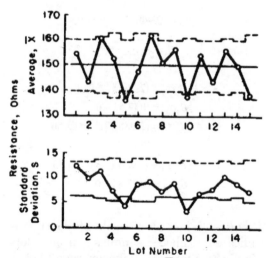

FIG. 19—Control charts for $\bar{X}$ and *s*. Small samples of unequal size, n = 3, 4, 5; μ_0, σ_0 given.

Central Lines
For $\bar{X}$: $\mu_0 = 150$

$$n = 3$$
$$\mu_0 \pm A\sigma_0 = 150 \pm 1.732(7.5)$$
137.0 and 163.0

$$n = 4$$
$$\mu_0 \pm A\sigma_0 = 150 \pm 1.500(7.5)$$
138.8 and 161.2

$$n = 5$$
$$\mu_0 \pm A\sigma_0 = 150 \pm 1.342(7.5)$$
139.9 and 160.1

TABLE 39. Resistance in ohms after 100-h operation, lot by lot control data.

Sample	Sample Size n	Average $\overline{X}$	Standard Deviation s	Sample	Sample Size n	Average $\overline{X}$	Standard Deviation s
1	5	154.6	12.20	9	5	156.2	8.92
2	5	143.4	9.75	10	4	137.5	3.24
3	4	160.8	11.20	11	5	153.8	6.85
4	3	152.7	7.43	12	5	143.4	7.64
5	5	136.0	4.32	13	4	156.0	10.18
6	3	147.3	8.65	14	5	149.8	8.86
7	3	161.7	9.23	15	3	138.2	7.38
8	5	151.0	7.24				

For s: $\sigma_0 = 7.5$

$$n = 3$$
$$c_4\sigma_0 = (0.8862)(7.5) = 6.65$$
$$n = 4$$
$$c_4\sigma_0 = (0.9213)(7.5) = 6.91$$
$$n = 5$$
$$c_4\sigma_0 = (0.9400)(7.5) = 7.05$$

For s: $\sigma_0 = 7.5$

$$n = 3 : B_6\sigma_0 = (2.276)(7.5) = 17.07$$
$$B_5\sigma_0 = (0)(7.5) = 0$$
$$n = 4 : B_6\sigma_0 = (2.088)(7.5) = 15.66$$
$$B_5\sigma_0 = (0)(7.5) = 0$$
$$n = 5 : B_6\sigma_0 = (1.964)(7.5) = 14.73$$
$$B_5\sigma_0 = (0)(7.5) = 0$$

Results—Evidence of lack of control is indicated since samples from lots Numbers 5 and 10 have averages below their lower control limit. No standard deviation values are outside their control limits. Corrective action is required to reduce the variation between lot averages.

Example 16: Control Charts for $\overline{X}$ and R, Small Samples of Equal Size (Section 19 and 20)

Consider the same problem as in *Example 12* where $\mu_0 = 35.00$ lb and $\sigma_0 = 4.20$ lb. The manufacturer wished to control variations in quality from lot to lot by taking a small sample from each lot. Table 40 gives observed values of $\overline{X}$ and R for samples of n = 5 each, selected from ten consecutive lots. Since the sample size n is less than 10, actually 5, he elected to use control charts for $\overline{X}$ and R rather than for $\overline{X}$ and s. Figure 20 gives the control charts for $\overline{X}$ and R.

TABLE 40. Operating characteristic, lot by lot control data.

Lot	Sample Size n	Average $\overline{X}$	Range R
No. 1	5	36.0	6.6
No. 2	5	31.4	0.5
No. 3	5	39.0	15.1
No. 4	5	35.6	8.8
No. 5	5	38.8	2.2
No. 6	5	41.6	3.5
No. 7	5	36.2	9.6
No. 8	5	38.0	9.0
No. 9	5	31.4	20.6
No. 10	5	29.2	21.7

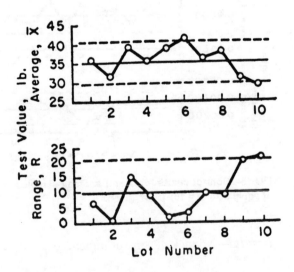

FIG. 20—Control charts for $\overline{X}$ and *R*. Small samples of equal size, *n* = 5; μ_0, σ_0 given.

Central Lines

For $\bar{X}$: $\mu_0 = 35.00$

$n = 5$

For R: $d_2\sigma_0 = 2.326(4.20) = 9.8$

Control Limits

$n = 5$

For $\bar{X}$: $\mu_0 \pm A\sigma_0 =$

$35.00 \pm (1.342)(4.20)$

29.4 and 40.6

For R: $d_2\sigma_0 = (4.918)(4.20) = 20.7$

$A_1\sigma_0 = (0)(4.20) = 0$

Results—Lack of control at the standard level is indicated by results for lot numbers 6 and 10. Corrective action is required both with respect to averages and with respect to variability within a lot.

Example 17: Control Charts for p, Samples of Equal Size (Section 23) and np, Samples of Equal Size (Section 24)

Consider the same data as in *Example 7*, Table 31. The manufacturer wishes to control his process with respect to finish on galvanized washers at a level such that the fraction nonconforming $p_0 = 0.0040$ (4 nonconforming washers per thousand). Table 31 of *Example 7* gives observed values of "number of nonconforming units" for finish nonconformities such as exposed steel, rough galvanizing in samples of 400 washers drawn from 15 successive lots. Figure 21 shows the control chart for p, and Fig. 22 gives the control chart for np. In practice, only one of these control charts would be used since, except for change of scale, the two charts are identical.

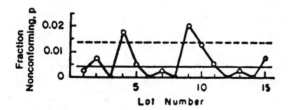

FIG. 21—Control chart for *p*. Samples of equal size, *n* = 400; *p₀* given.

(a) *p*

Central Line

$p_0 = 0.0040$

Control Limits

$n = 400$

$$p_0 \pm 3\sqrt{\frac{p_0(1-p_0)}{n}} =$$

$$0.0040 \pm 3\sqrt{\frac{0.0040(0.9960)}{400}} =$$

0.0040 ± 0.0095

0 and 0.0135

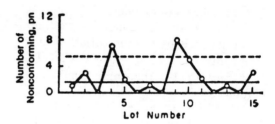

FIG. 22—Control chart for *np*. Samples of equal size, *n* = 400; *p₀* given.

(b) *np*

Central Line

$np_0 = 0.0040(400) = 1.6$

Control limits

Extract formula:

$n = 400$

$$np_0 \pm 3\sqrt{np_0(1-p_0)} =$$

$$1.6 \pm 3\sqrt{1.6(0.996)} =$$

$$1.6 \pm 3\sqrt{1.5936} =$$

$1.6 \pm 3(1.262)$

0 and 5.4

Simplified approximate formula:

$n = 400$

Since p_0 is small, replace Eq 29 by Eq 30

$$np_0 \pm 3\sqrt{np_0} =$$

$$1.6 \pm 3\sqrt{1.6} =$$

$1.6 \pm 3(1.265)$

0 and 5.4

Results—Lack of control of quality is indicated with respect to the desired level; lot numbers 4 and 9 are outside control limits.

NOTE

Since the value of np_0 is 1.6, less than 4, the **NOTE** at the end of Section 13 (or 14) applies, as mentioned at the end of Section 23 (or 24). The product of n and the upper control limit value for p is $400 \times 0.0135 = 5.4$. The nonintegral remainder, 0.4, is less than one-half. The upper control limit stands as does the indication of lack of control to p_0. For np, by the **NOTE** of Section 14, the same conclusion follows.

Example 18: Control Chart for p (Fraction Nonconforming), Samples of Unequal Size (Section 23e)

The manufacturer wished to control the quality of a type of electrical apparatus with respect to two adjustment characteristics at a level such that the fraction nonconforming $p_0 = 0.0020$ (2 nonconforming units per thousand). Table 41 gives observed values of "number of nonconforming units" for this item found in samples drawn from successive lots.

Sample sizes vary considerably from lot to lot and, hence, control limits are computed for each sample. Equivalent control limits for "number of nonconforming units," np, are shown in column 5 of the table. In this way, the original records showing "number of nonconforming units" may be compared directly with control limits for np. Figure 23 shows the control chart for p.

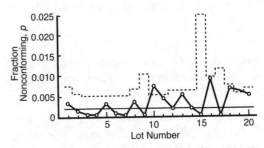

FIG. 23—Control chart for p. Samples of unequal size, to 2500; p_0 given.

TABLE 41. Adjustment irregularities, electrical apparatus.

LOT	SAMPLE SIZE n	NUMBER OF NON-CONFORMING UNITS	FRACTION NON-CONFORMING p	UPPER CONTROL LIMIT FOR np	UPPER CONTROL LIMIT FOR p
No. 1	600	2	0.0033	4.5	0.0075
No. 2	1300	2	0.0015	7.4	0.0057
No. 3	2000	1	0.0005	10.0	0.0050
No. 4	2500	1	0.0004	11.7	0.0047
No. 5	1550	5	0.0032	8.4	0.0054
No. 6	2000	2	0.0010	10.0	0.0050
No. 7	1550	0	0.0000	8.4	0.0054
No. 8	780	3	0.0038	5.3	0.0068
No. 9	260	0	0.0000	2.7	0.0103
No. 10	2000	15	0.0075	10.0	0.0050
No. 11	1550	7	0.0045	8.4	0.0054
No. 12	950	2	0.0021	6.0	0.0063
No. 13	950	5	0.0053	6.0	0.0063
No. 14	950	2	0.0021	6.0	0.0063
No. 15	35	0	0.0000	0.9	0.0247
No. 16	330	3	0.0091	3.1	0.0094
No. 17	200	0	0.0000	2.3	0.0115
No. 18	600	4	0.0067	4.5	0.0075
No. 19	1300	8	0.0062	7.4	0.0057
No. 20	780	4	0.0051	5.3	0.0068

Central Line for p

$$p_0 = 0.0020$$

Control Limits for p

$$p_0 \pm 3\sqrt{\frac{p_0(1-p_0)}{n}}$$

For $n = 600$:

$$0.0020 \pm 3\sqrt{\frac{0.002(0.998)}{600}} =$$

$$0.0020 \pm 3(0.001824)$$

$$0 \text{ and } 0.0075$$

(same procedure for other values of n)

Control Limits for np

Using Eq 30 for np,

$$np_0 \pm 3\sqrt{np_0}$$

For $n = 600$:

$$1.2 \pm 3\sqrt{1.2} = 1.2 \pm 3(1.095),$$

$$0 \text{ and } 4.5$$

(same procedure for other values of n)

Results—Lack of control and need for corrective action indicated by results for lots Numbers 10 and 19.

NOTE

The values of np_0 for these lots are 4.0 and 2.6, respectively. The **NOTE** at the end of Section 13 (or 14) applies to lot number 19. The product of n and the upper control limit value for p is $1300 \times 0.0057 = 7.41$. The nonintegral remainder is 0.41, less than one-half. The upper control limit stands, as does the indication of lack of control at p_0. For np, by the **NOTE** of Section 14, the same conclusion follows.

Example 19: Control Chart for p (Fraction Rejected), Total and Components, Samples of Unequal Size (Section 23)

A control device was given a 100 percent inspection in lots varying in size from about 1800 to 5000 units, each unit being tested and inspected with respect to 23 *essentially independent* characteristics. These 23 characteristics were grouped into three groups designated Groups A, B, and C, corresponding to three successive inspections.

A unit found nonconforming at any time with respect to any one characteristic was immediately rejected; hence units found nonconforming in, say, the Group A inspection were not subjected to the two subsequent group inspections. In fact, the number of units inspected for each characteristic in a group itself will differ from characteristic to characteristic if nonconformities with respect to the characteristics in a group occur, the last characteristic in the group having the smallest sample size.

Since 100 percent inspection is used, no additional units are available for inspection to maintain a constant sample size for all characteristics in a group or for all the component groups. The fraction nonconforming with respect to each characteristic is sufficiently small so that the error within a group, although rather large between the first and last characteristic inspected by one inspection group, can be neglected for practical purposes. Under these circumstances, the number inspected for any group was equal to the lot size diminished by the number of units rejected in the preceding inspections.

Part 1 of Table 42 gives the data for 12 successive lots of product, and shows for each lot inspected the total fraction rejected as well as the number and fraction rejected at each inspection station. Part 2 of Table 42 gives values of

p_0, fraction rejected, at which levels the manufacturer desires to control this device, with respect to all twenty-three characteristics combined and with respect to the characteristics tested and inspected at each of the three inspection stations. Note that the p_0 for all characteristics (in terms of nonconforming units) is less than the sum of the p_0 values for the three component groups, since nonconformities from more than one characteristic or group of characteristics may occur on a single unit. Control limits, lower and upper, in terms of fraction rejected are listed for each lot size using the initial lot size as the sample size for all characteristics combined and the lot size available at the beginning of inspection and test for each group as the sample size for that group.

TABLE 42. Inspection data for 100 percent inspection—control device.

	\multicolumn{12}{c}{Observed Number of Rejects and Fraction Rejected}											
	All Groups Combined			Group A			Group B			Group C		
	Lot	Total Rejected		Lot	Rejected		Lot	Rejected		Lot	Rejected	
Lot	Size, n	Number	Fraction	Size, n	Number	Fraction	Size, n	Number	Fraction	Size, n	Number	Fraction
No. 1	4814	914	0.190	4814	311	0.065	4503	253	0.056	4250	350	0.082
No .2	2159	359	0.166	2159	128	0.059	2031	105	0.052	1926	126	0.065
No. 3	3089	565	0.183	3089	195	0.063	2894	149	0.051	2745	221	0.081
No. 4	3156	626	0.198	3156	233	0.074	2923	142	0.049	2781	251	0.090
No. 5	2139	434	0.203	2139	146	0.068	1993	101	0.051	1892	187	0.099
No. 6	2588	503	0.194	2588	177	0.068	2411	151	0.063	2260	175	0.077
No. 7	2510	487	0.194	2510	143	0.057	2367	116	0.049	2251	228	0.101
No. 8	4103	803	0.196	4103	318	0.078	3785	242	0.064	3543	243	0.069
No. 9	2992	547	0.183	2992	208	0.070	2784	130	0.047	2654	209	0.079
No.10	3545	643	0.181	3545	172	0.049	3373	180	0.053	3193	291	0.091
No.11	1841	353	0.192	1841	97	0.053	1744	119	0.068	1625	137	0.084
No.12	2748	418	0.152	2748	141	0.051	2607	114	0.044	2493	163	0.065

\multicolumn{5}{c}{Central Lines and Control Limits, Based on Standard p_0 Values}				
	All Groups Combined	Group A	Group B	Group C
\multicolumn{5}{c}{Central Lines}				
$p_0=$	0.180	0.070	0.050	0.080
Lot	\multicolumn{4}{c}{Control Limits}			
No. 1	0.197 and 0.163	0.081 and 0.059	0.060 and 0.040	0.093 and 0.067
No. 2	0.205 and 0.155	0.086 and 0.054	0.064 and 0.036	0.099 and 0.061
No. 3	0.201 and 0.159	0.084 and 0.056	0.062 and 0.038	0.096 and 0.064
No. 4	0.200 and 0.160	0.084 and 0.056	0.062 and 0.038	0.095 and 0.065
No. 5	0.205 and 0.155	0.086 and 0.054	0.065 and 0.035	0.099 and 0.061
No. 6	0.203 and 0.157	0.085 and 0.055	0.063 and 0.037	0.097 and 0.063
No. 7	0.203 and 0.157	0.085 and 0.055	0.064 and 0.036	0.097 and 0.063
No. 8	0.198 and 0.162	0.082 and 0.058	0.061 and 0.039	0.094 and 0.066
No. 9	0.201 and 0.159	0.084 and 0.056	0.062 and 0.038	0.096 and 0.064
No. 10	0.200 and 0.160	0.083 and 0.057	0.061 and 0.039	0.094 and 0.066
No. 11	0.207 and 0.153	0.088 and 0.052	0.066 and 0.034	0.100 and 0.060
No. 12	0.202 and 0.158	0.085 and 0.055	0.063 and 0.037	0.096 and 0.064

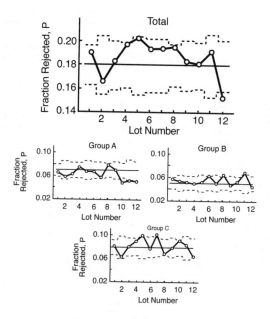

FIG. 24—Control charts for *p* (fraction rejected) for total and components. Samples of unequal size, *n* = 1625 to 4814; p_0 given.

Figure 24 shows four control charts, one covering all rejections combined for the control device and three other charts covering the rejections for each of the three inspection stations for Group A, Group B, and Group C characteristics, respectively. Detailed computations for the over-all results for one lot and one of its component groups are given.

Central Lines

See Table 42

Control Limits

See Table 42
For Lot Number 1

Total: $n = 4814$

$$p_0 \pm 3\sqrt{\frac{p_0(1-p_0)}{n}} =$$

$$0.180 \pm 3\sqrt{\frac{0.180(0.820)}{4814}} =$$

$$0.180 \pm 3(0.0055)$$

$$0.163 \text{ and } 0.197$$

Group C: $n = 4250$

$$p_0 \pm 3\sqrt{\frac{p_0(1-p_0)}{n}} =$$

$$0.080 \pm 3\sqrt{\frac{0.080(0.920)}{4250}} =$$

$$0.080 \pm 3(0.0042)$$

$$0.067 \text{ and } 0.093$$

Results—Lack of control is indicated for all characteristics combined; lot number 12 is outside control limits in a favorable direction and the corresponding results for each of the three components are less than their standard values, Group A being below the lower control limit. For Group A results, lack of control is indicated since lot numbers 10 and 12 are below their lower control limits. Lack of control is indicated for the component characteristics in Group B, since lot numbers 8 and 11 are above their upper control limits. For Group C, lot number 7 is above its upper limit indicating lack of control. Corrective measures are indicated for Groups B and C and steps should be taken to determine whether the Group A component might not be controlled at a smaller value of p_0, such as 0.06. The values of np_0 for lot numbers 8 and 11 in Group B and lot number 7 in Group C are all larger than 4. The **NOTE** at the end of Section 13 does not apply.

Example 20: Control Chart for u, Samples of Unequal Size (Section 25)

It is desired to control the number of nonconformities per billet to a standard of 1.000 nonconformities per unit in order that the wire made from such billets of copper will not contain an excessive number of nonconformities. The lot sizes varied greatly from day to day so that a sampling schedule was set up giving three different samples sizes to cover the range of lot sizes received. A control program was instituted using a control chart for

TABLE 43. Lot by lot inspection results for copper billets in terms of number of nonconformities and nonconformities per unit.

LOT	SAMPLE SIZE n	NUMBER OF NON-CONFORMITIES c	NON-CONFORMITIES PER UNIT u	LOT	SAMPLE SIZE n	NUMBER OF NONCONFORMITIES c	NON-CONFORMITIES PER UNIT u
No. 1	100	75	0.750	No. 10	100	130	1.300
No. 2	100	138	1.380	No. 11	100	58	0.580
No. 3	200	212	1.060	No. 12	400	480	1.200
No. 4	400	444	1.110	No. 13	400	316	0.790
No. 5	400	508	1.270	No. 14	200	162	0.810
No. 6	400	312	0.780	No. 15	200	178	0.890
No. 7	200	168	0.840				
No. 8	200	266	1.330	Total	3500	3566	
No. 9	100	119	1.190	Overall[a]			1.019

[a] $\bar{u} = 3566/3500 = 1.019$.

nonconformities per unit with reference to the desired standard. Table 43 gives data in terms of nonconformities and nonconformities per unit for 15 consecutive lots under this program. Figure 25 shows the control chart for u.

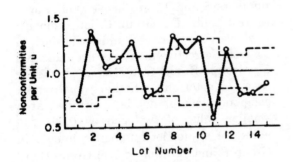

FIG. 25—Control chart for u. Samples of unequal size, n = 100, 200, 400; u_0 given.

Central Line
$u_0 = 1.000$

Control Limits
$n = 100$

$$u_0 \pm 3\sqrt{\frac{u_0}{n}} =$$

$$1.000 \pm 3\sqrt{\frac{1.000}{100}} =$$

$$1.000 \pm 3(0.100)$$

0.700 and 1.300

$n = 200$

$$u_0 \pm 3\sqrt{\frac{u_0}{n}} =$$

$$1.000 \pm 3\sqrt{\frac{1.000}{200}} =$$

$$1.000 \pm 3(0.0707)$$

0.788 and 1.212

$n = 400$

$$u_0 \pm 3\sqrt{\frac{u_0}{n}} =$$

$$1.000 \pm 3\sqrt{\frac{1.000}{400}} =$$

$$1.000 \pm 3(0.0500)$$

0.850 and 1.150

Results—Lack of control of quality is indicated with respect to the desired level since lot numbers 2, 5, 8, and 12 are above the upper control limit and lot numbers 6, 11, and 13 are below the lower control limit. The overall level, 1.019 nonconformities per unit, is slightly above the desired value of 1.000 nonconformities per unit. Corrective action is necessary to reduce the spread between successive lots and reduce the average number of nonconformities per unit. The values of np_0 for all lots are at least 100 so that the **NOTE** at end of Section 15 does not apply.

TABLE 44. Daily inspection results for Type D motors in terms of nonconformities per sample and nonconformities per unit.

LOT	SAMPLE SIZE n	NUMBER OF NONCONFORMITIES c	NONCONFORMITIES PER UNIT u
No. 1	25	81	3.24
No. 2	25	64	2.56
No. 3	25	53	2.12
No. 4	25	95	3.80
No. 5	25	50	2.00
No. 6	25	73	2.92
No. 7	25	91	3.64
No. 8	25	86	3.44
No. 9	25	99	3.96
No. 10	25	60	2.40
Total	250	752	30.08
Average	25.0	75.2	3.008

Example 21: Control Charts for c, Samples of Equal Size (Section 26)

A Type D motor is being produced by a manufacturer that desires to control the number of nonconformities per motor at a level of $u_0 = 3.000$ nonconformities per unit with respect to all visual nonconformities. The manufacturer produces on a continuous basis and decides to take a sample of 25 motors every day, where a day's product is treated as a lot. Because of the nature of the process, plans are to control the product for these nonconformities at a level such that $c_0 = 75.0$ nonconformities and $nu_0 = c_0$. Table 44 gives data in terms of number of nonconformities, *c*, and also the number of nonconformities per unit, *u*, for 10 consecutive days. Figure 26 shows the control chart for *c*. As in *Example 20*, a control chart may be made for *u*, where the central line is $u_0 = 3.000$ and the control limits are

$$u_0 \pm 3\sqrt{\frac{u_0}{n}} =$$

$$3.000 \pm 3\sqrt{\frac{3.000}{25}} =$$

$$3.000 \pm 3(0.3464)$$

$$1.96 \text{ and } 4.04$$

Central Line

$$c_0 = nu_0 = 3.000 \times 25 = 75.0$$

Control Limits

$$n = 25$$

$$c_0 \pm 3\sqrt{c_0} =$$

$$75.0 \pm 3\sqrt{75.0} =$$

$$75.0 \pm 3(8.66)$$

$$49.02 \text{ and } 100.98$$

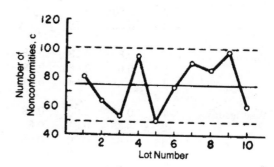

FIG. 26—Control chart for c. Sample of equal size, *n* = 25; c_0 given.

Results—No significant deviations from the desired level. There are no points outside limits so that the **NOTE** at the end of Section 16 does not apply. In addition, $c_0 = 75$, larger than 4.

33. Illustrative Examples—Control Chart for Individuals

Examples 22 to 25, inclusive, illustrate the use of the control chart for individuals, in which individual observations are plotted one by one. The examples cover the two general conditions: (*a*) control, no standard given; and (*b*) control with respect to a given standard (see Sections 28 to 30).

Example 22: Control Chart for Individuals, X — Using Rational Subgroups, Samples of Equal Size, No Standard Given — Based on $\overline{\overline{X}}$ and $\overline{MR}$ (Section 29)

In the manufacture of manganese steel tank shoes, five 4-ton heats of metal were cast in each 8-h shift, the silicon content being controlled by ladle additions computed from preliminary analyses.

High silicon content was known to aid in the production of sound castings, but the specification set a maximum of 1.00 percent silicon for a heat, and all shoes from a heat exceeding this specification were rejected. It was important, therefore, to detect any trouble with silicon control before even one heat exceeded the specification.

Since the heats of metal were well stirred, within-heat variation of silicon content was not a useful basis for control limits. However, each 8-h shift used the same materials, equipment etc., and the quality depended largely on the care and efficiency with which they operated so that the five heats produced in an 8-h shift provided a rational subgroup.

Data analyzed in the course of an investigation and before standard values were established are shown in Table 45

TABLE 45. Silicon content of heats of manganese steel, percent.

Day	Shift	Heat 1	2	3	4	5	Size, n	Sample Average, $\overline{X}$	Range, R
Monday	1	0.70	0.72	0.61	0.75	0.73	5	0.702	0.14
	2	0.83	0.68	0.83	0.71	0.73	5	0.756	0.15
	3	0.86	0.78	0.71	0.70	0.90	5	0.790	0.20
Tuesday	1	0.80	0.78	0.68	0.70	0.74	5	0.740	0.12
	2	0.64	0.66	0.79	0.81	0.68	5	0.716	0.17
	3	0.68	0.64	0.71	0.69	0.81	5	0.706	0.17
Wednesday	1	0.80	0.63	0.69	0.62	0.75	5	0.698	0.18
	2	0.65	0.81	0.68	0.84	0.66	5	0.728	0.19
	3	0.64	0.70	0.66	0.65	0.93	5	0.716	0.29
Thursday	1	0.77	0.83	0.88	0.70	0.64	5	0.764	0.24
	2	0.72	0.67	0.77	0.74	0.72	5	0.724	0.10
	3	0.73	0.66	0.72	0.73	0.71	5	0.710	0.07
Friday	1	0.79	0.70	0.63	0.70	0.88	5	0.740	0.25
	2	0.85	0.80	0.78	0.85	0.62	5	0.780	0.23
	3	0.67	0.78	0.81	0.84	0.96	5	0.812	0.29
Total	15							11.082	2.79
Average								0.7388	0.186

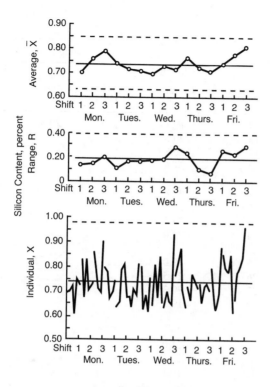

FIG. 27—Control charts for $\overline{X}$, R, and X. Samples of equal size, $n = 5$; no standard given.

and control charts for $\overline{X}$, MR, and X are shown in Fig. 27.

Central Lines

For $\overline{X}$: $\overline{\overline{X}} = 0.7388$

For R: $\overline{R} = 0.186$

For X: $\overline{\overline{X}} = 0.7388$

Control Limits
$n = 5$

For $\overline{X}$: $\overline{\overline{X}} \pm A_2 \overline{R} =$
$0.7388 \pm (0.577)(0.186)$
0.631 and 0.846

For R: $D_4 \overline{R} = (2.115)(0.186) = 0.393$
$D_3 \overline{R} = (0)(0.186) = 0$

For X: $\overline{\overline{X}} \pm E_2 \overline{MR} =$
$0.7388 \pm (1.290)(0.186)$
0.499 and 0.979

Results—None of the charts give evidence of lack of control.

Example 23: Control Chart for Individuals, X—Using Rational Subgroups, Standard Given, Based on μ_0 and σ_0 (Section 29)

In the hand spraying of small instrument pins held in bar frames of 25 each, coating thickness and weight had to be delicately controlled and spray-gun adjustments were critical and had to be watched continuously from bar to bar. Weights were measured by careful weighing before and after removal of the coating. Destroying more than one pin per bar was economically not feasible, yet failure to catch a bar departing from standards might result in the unsatisfactory performance of some 24 assembled instruments. The standard lot size for these instrument pins was 100 so that initially control charts for average and range were set up with $n = 4$. It was found that the variation in thickness of coating on the 25 pins on a single bar was quite small as compared with the between-bar variation. Accordingly, as an adjunct to the control charts for average and range, a control chart for individuals, X, at the sprayer position was adopted for the operator's guidance.

Table 46 gives data comprising observations on 32 pins taken from consecutive bar frames together with 8 average and range values where $n = 4$. It was desired to control the weight with an average $\mu_0 = 20.00$ mg and $\sigma_0 = 0.900$ mg. Figure 28 shows the control chart for individual values X for coating weights of instrument pins together with the control charts for $\overline{X}$ and R for samples where $n = 4$.

TABLE 46. Coating weights of instrument pins, milligrams.

Individual	Individual Observation, X	Sample, $n = 4$			Individual	Individual Observation, X	Sample, $n = 4$		
		Sample	Average, $\overline{X}$	Range, R			Sample	Average, $\overline{X}$	Range, R
1	18.5	1	18.90	4.7	17	19.1	5	20.52	2.5
2	21.2				18	20.6			
3	19.4				19	20.8			
4	16.5				20	21.6			
5	17.9	2	19.60	3.3	21	22.8	6	22.80	1.0
6	19.0				22	22.2			
7	20.3				23	23.2			
8	21.2				24	23.0			
9	19.6	3	20.08	0.9	25	19.0	7	19.75	1.5
10	19.8				26	20.5			
11	20.4				27	20.3			
12	20.5				28	19.2			
13	22.2	4	21.20	1.9	29	20.7	8	20.32	1.9
14	21.5				30	21.0			
15	20.8				31	20.5			
16	20.3				32	19.1			
					Total	652.7		163.17	17.7
					Average	20.40		20.40	2.21

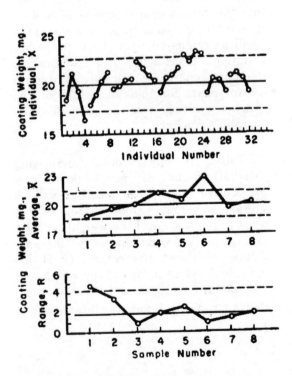

FIG. 28—Control charts for X, $\overline{X}$, and R. Small samples of equal size, $n = 4$; μ_0, σ_0 given.

Central Line

For X : $\mu_0 = 20.00$

Control Limits

For X : $\mu_0 \pm 3\sigma_0 =$
$20.00 \pm 3(0.900)$
17.3 and 22.7

Central Lines

For $\overline{X}$: $\mu_0 = 20.00$

For R: $d_2\sigma_0 = (2.059)(0.900) = 1.85$

Control Limits

$n = 4$

For $\overline{X}$: $\mu_0 \pm A\sigma_0 =$
$20.00 \pm (1.500)(0.900)$
18.65 and 21.35

For R: $D_2\sigma_0 = (4.698)(0.900) = 4.23$

$D_1\sigma_0 = (0)(0.900) = 0$

Results—All three charts show lack of control. At the outset, both the chart for ranges and the chart for individuals gave indications of lack of control. Subsequently, for Sample 6 the control chart for individuals showed the first unit in the sample of 4 to be outside its upper control limit, thus indicating lack of control before the entire sample was obtained.

Example 24: Control Charts for Individuals, X, and Moving Range, MR, of Two Observations, No Standard Given— Based on $\overline{X}$ and $\overline{MR}$ the Mean Moving Range (Section 30A)

A distilling plant was distilling and blending batch lots of denatured alcohol in a large tank. It was desired to control the percentage of methanol for this process. The variability of sampling within a single lot was found to be negligible so it was decided feasible to take only one observation per lot and to set control limits based on the moving range of successive lots. Table 47 gives a summary of the methanol content, X, of 26 consecutive lots of the denatured alcohol and the 25 values of the moving range, MR, the range of successive lots with $n = 2$. Figure 29 gives control charts for individuals, X, and the moving range, MR.

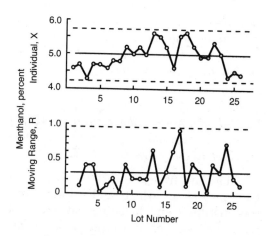

Fig. 29—Control charts for X and MR. No standard given; based on moving range, where n = 2.

Central Lines

For X : $\overline{X} = \dfrac{128.1}{26} = 4.927$

For R : $\overline{R} = \dfrac{7.2}{25} = 0.288$

Control Limits

$n = 2$

For X: $\overline{\overline{X}} \pm E_2 \overline{MR} = \overline{\overline{X}} \pm 2.660\,\overline{MR} =$

$4.927 \pm (2.660)(0.288)$

4.2 and 5.7

For R : $D_4 \overline{MR} = (3.267)(0.288) = 0.94$

$D_3 \overline{MR} = (0)(0.288) = 0$

TABLE 47. Methanol content of successive lots of denatured alcohol and moving range for $n = 2$.

LOT	PERCENTAGE OF METHANOL X	MOVING RANGE MR	LOT	PERCENTAGE OF METHANOL X	MOVING RANGE MR
No. 1	4.6	...	No. 14	5.5	0.1
No. 2	4.7	0.1	No. 15	5.2	0.3
No. 3	4.3	0.4	No. 16	4.6	0.6
No. 4	4.7	0.4	No. 17	5.5	0.9
No. 5	4.7	0	No. 18	5.6	0.1
No. 6	4.6	0.1	No. 19	5.2	0.4
No. 7	4.8	0.2	No. 20	4.9	0.3
No. 8	4.8	0	No. 21	4.9	0
No. 9	5.2	0.4	No. 22	5.3	0.4
No. 10	5.0	0.2	No. 23	5.0	0.3
No. 11	5.2	0.2	No. 24	4.3	0.7
No. 12	5.0	0.2	No. 25	4.5	0.2
No. 13	5.6	0.6	No. 26	4.4	0.1
			Total	128.1	7.2

Results—The trend pattern of the individuals and their tendency to crowd the control limits suggests that better control may be attainable.

Example 25: Control Charts for Individuals, X, and Moving Range, MR, of Two Observations, Standard Given—Based on μ_0 and σ_0 (Section 30B)

The data are from the same source as for *Example 24*, in which a distilling plant was distilling and blending batch lots of denatured alcohol in a large tank. It was desired to control the percentage of water for this process. The variability of sampling within a single lot was found to be negligible so it was decided to take only one observation per lot and to set control limits for individual values, X, and for the moving range, *MR*, of successive lots with n = 2 where μ_0 = 7.800 percent and σ_0 = 0.200 percent. Table 48 gives a summary of the water content of 26 consecutive lots of the denatured alcohol and the 25 values of the moving range, R. Figure 30 gives control charts for individuals, X, and for the moving range, *MR*.

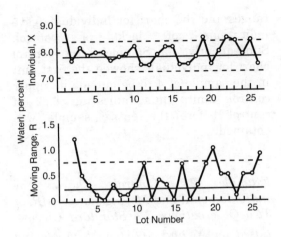

Fig. 30—Control charts for *X* and moving range, *MR*, where *n* = 2. Standard given; based on μ_0 and σ_0.

Central Lines

For X: $\mu_0 = 7.800$

$n = 2$

For R: $d_2\sigma_0 = (1.128)(0.200) = 0.23$

Control Limits

For *X*: $\mu_0 \pm 3\sigma =$

$7.800 \pm 3(0.200)$

7.2 and 8.4

$n = 2$

For *R*: $D_2\sigma_0 = (3.686)(0.200) = 0.74$

$D_1\sigma_0 = (0)(0.200) = 0$

TABLE 48. Water content of successive lots of denatured alcohol and moving range for *n* = 2.

Lot	Percentage of Water, X	Moving Range, MR	Lot	Percentage of Water, X	Moving Range, MR
No. 1	8.9	...	No. 14	8.2	0.3
No. 2	7.7	1.2	No. 15	8.2	0
No. 3	8.2	0.5	No. 16	7.5	0.7
No. 4	7.9	0.3	No. 17	7.5	0
No. 5	8.0	0.1	No. 18	7.8	0.3
No. 6	8.0	0	No. 19	8.5	0.7
No. 7	7.7	0.3	No. 20	7.5	1.0
No. 8	7.8	0.1	No. 21	8.0	0.5
No. 9	7.9	0.1	No. 22	8.5	0.5
No. 10	8.2	0.3	No. 23	8.4	0.1
No. 11	7.5	0.7	No. 24	7.9	0.5
No. 12	7.5	0	No. 25	8.4	0.5
No. 13	7.9	0.4	No. 26	7.5	0.9
			Total	207.1	10.0
			Number of values	26	25
			Average	7.965	0.400

Results—Lack of control at desired levels is indicated with respect to both the individual readings and the moving range. These results indicate corrective measures should be taken to reduce the level in percent and to reduce the variation between lots.

SUPPLEMENT A

Mathematical Relations and Tables of Factors for Computing Control Chart Lines

Scope

Supplement A presents mathematical relations used in arriving at the factors and formulas of **PART 3**. In addition, **Supplement A** presents approximations to c_4, $1/c_4$, B_3, B_4, B_5, and B_6 for use when needed. Finally, a more comprehensive tabulation of values of these factors is given in Tables 49 and 50, including reciprocal values of c_4 and d_2, and values of d_3.

Factors c_4, d_2, and d_3, (values for $n = 2$ to 25, inclusive, in Table 49)

The relations given for factors c_4, d_2, and d_3 are based on sampling from a universe having a Normal distribution (see Ref. **1**, p. 184).

$$c_4 = \sqrt{\frac{2}{n-1}} \frac{\left(\frac{n-2}{2}\right)!}{\left(\frac{n-3}{2}\right)!} \qquad (42)$$

where the symbol $(k/2)!$ is called "$k/2$ factorial" and satisfies the relations $(-1/2)! = \sqrt{\pi}$, $0! = 1$, and $(k/2)! = (k/2)[((k-2)/2)!]$ for $k = 1, 2, 3, \ldots$. If k is even, $(k/2)!$ is simply the product of all integers from $k/2$ down to 1; for example, if $k = 8$, $(8/2)! = 4!$

$= 4 \cdot 3 \cdot 2 \cdot 1 = 24$. If k is odd, $(k/2)!$ is the product of all half-integers from $k/2$ down to 1/2, multiplied by $\sqrt{\pi}$; for example, if $k = 7$, so $(7/2)! = (7/2) \cdot (5/2) \cdot (3/2) \cdot (1/2) \cdot \sqrt{\pi} = 11.6317$.

$$d_2 = \int_{-\infty}^{\infty} \left[1 - (1 - \alpha_1)^n - \alpha_1^n\right] dx_1 \qquad (43)$$

where

$$\alpha_1 = \frac{1}{\sqrt{2\pi}} \int_{-\infty}^{x_1} e^{-(x^2/2)} dx, \text{ and } n = \text{sample size.}$$

$$d_3 = \sqrt{2\int_{-\infty}^{\infty}\int_{-\infty}^{x_1}[1 - \alpha_1^n - (1-\alpha_n)^n + (\alpha_1 - \alpha_n)^n]dx_n dx_1 - d_2^2}$$

$$(44)$$

where

$$\alpha_1 = \frac{1}{\sqrt{2\pi}} \int_{-\infty}^{x_1} e^{-(x^2/2)} dx$$

$$\alpha_n = \frac{1}{\sqrt{2\pi}} \int_{-\infty}^{x_n} e^{-(x^2/2)} dx$$

n = sample size, and d_2 = average range for a Normal law distribution with standard deviation equal to unity. (In his original paper, Tippett [9] used w for the range and $\overline{w}$ for d_2.)

The relations just mentioned for c_4, d_2, and d_3 are exact when the original universe is Normal but this does not limit their use in practice. They may for most practical purposes be considered satisfactory for use in control chart work although the universe is not Normal. Since the relations are involved and thus difficult to compute, values of c_4, d_2, and d_3 for $n = 2$ to 25, inclusive, are given in Table 49. All values listed in the table were computed to enough significant figures so that when rounded off in accordance with standard practices the last figure shown in the table was not in doubt.

Standard Deviations of $\overline{X}$, s, R, p, np, u, *and* c

The standard deviations of $\overline{X}$, s, R, p, etc., used in setting 3-sigma control limits and designated $\sigma_{\overline{x}}$, σ_s, σ_R, σ_p, etc., in **PART 3**, are the standard deviations of the sampling distributions of $\overline{X}$, s, R, p, etc., for subgroups (samples) of size n. They are not the standard deviations which might be computed from the subgroup values of $\overline{X}$, s, R, p, etc., plotted on the control charts but are computed by formula from the quantities listed in Table 51.

TABLE 49. Factors for computing control chart lines.

OBSERVA-TIONS IN SAMPLE, n	CHART FOR AVERAGES FACTORS FOR CONTROL LIMITS			CHART FOR STANDARD DEVIATIONS FACTORS FOR CENTRAL LINE		FACTORS FOR CONTROL LIMITS				CHART FOR RANGES FACTORS FOR CENTRAL LINE		FACTORS FOR CONTROL LIMITS				
	A	A_2	A_3	c_4	$1/c_4$	B_3	B_4	B_5	B_6	d_2	$1/d_2$	d_3	D_1	D_2	D_3	D_4
2	2.121	1.880	2.659	0.7979	1.2533	0	3.267	0	2.606	1.128	0.8862	0.853	0	3.686	0	3.267
3	1.732	1.023	1.954	0.8862	1.1284	0	2.568	0	2.276	1.693	0.5908	0.888	0	4.358	0	2.575
4	1.500	0.729	1.628	0.9213	1.0854	0	2.266	0	2.088	2.059	0.4857	0.880	0	4.698	0	2.282
5	1.342	0.577	1.427	0.9400	1.0638	0	2.089	0	1.964	2.326	0.4299	0.864	0	4.918	0	2.114
6	1.225	0.483	1.287	0.9515	1.0510	0.030	1.970	0.029	1.874	2.534	0.3946	0.848	0	5.079	0	2.004
7	1.134	0.419	1.182	0.9594	1.0424	0.118	1.882	0.113	1.806	2.704	0.3698	0.833	0.205	5.204	0.076	1.924
8	1.061	0.373	1.099	0.9650	1.0363	0.185	1.815	0.179	1.751	2.847	0.3512	0.820	0.388	5.307	0.136	1.864
9	1.000	0.337	1.032	0.9693	1.0317	0.239	1.761	0.232	1.707	2.970	0.3367	0.808	0.547	5.393	0.184	1.816
10	0.949	0.308	0.975	0.9727	1.0281	0.284	1.716	0.276	1.669	3.078	0.3249	0.797	0.686	5.469	0.223	1.777
11	0.905	0.285	0.927	0.9754	1.0253	0.321	1.679	0.313	1.637	3.173	0.3152	0.787	0.811	5.535	0.256	1.744
12	0.866	0.266	0.886	0.9776	1.0230	0.354	1.646	0.346	1.610	3.258	0.3069	0.778	0.923	5.594	0.283	1.717
13	0.832	0.249	0.850	0.9794	1.0210	0.382	1.618	0.374	1.585	3.336	0.2998	0.770	1.025	5.647	0.307	1.693
14	0.802	0.235	0.817	0.9810	1.0194	0.406	1.594	0.399	1.563	3.407	0.2935	0.763	1.118	5.696	0.328	1.672
15	0.775	0.223	0.789	0.9823	1.0180	0.428	1.572	0.421	1.544	3.472	0.2880	0.756	1.203	5.782	0.363	1.637
16	0.750	0.212	0.763	0.9835	1.0168	0.448	1.552	0.440	1.526	3.532	0.2831	0.750	1.282	5.820	0.378	1.622
17	0.728	0.203	0.739	0.9845	1.0157	0.466	1.534	0.458	1.511	3.588	0.2787	0.744	1.356	5.856	0.391	1.609
18	0.707	0.194	0.718	0.9854	1.0148	0.482	1.518	0.475	1.496	3.640	0.2747	0.739	1.424	5.889	0.404	1.596
19	0.688	0.187	0.698	0.9862	1.0140	0.497	1.503	0.490	1.483	3.689	0.2711	0.733	1.489	5.921	0.415	1.585
20	0.671	0.180	0.680	0.9869	1.0132	0.510	1.490	0.504	1.470	3.735	0.2677	0.729	1.549	5.951	0.425	1.575
21	0.655	0.173	0.663	0.9876	1.0126	0.523	1.477	0.516	1.459	3.778	0.2647	0.724	1.606	5.979	0.435	1.565
22	0.640	0.167	0.647	0.9882	1.0120	0.534	1.466	0.528	1.448	3.819	0.2618	0.720	1.660	6.006	0.443	1.557
23	0.626	0.162	0.633	0.9887	1.0114	0.545	1.455	0.539	1.438	3.858	1.2592	0.716	1.711	6.032	0.452	1.548
24	0.612	0.157	0.619	0.9892	1.0109	0.555	1.445	0.549	1.429	3.895	0.2567	0.712	1.759	6.056	0.459	1.541
25	0.600	0.153	0.606	0.9896	1.0105	0.565	1.435	0.559	1.420	3.931	0.2544	0.708	1.805			
Over 25	$3/\sqrt{n}$	...	a	b	c	d	e	f	g	...	...	...	...	...	...	...

NOTES—Values of all factors in this Table were recomputed in 1987 by A. T. A. Holden of the Rochester Institute of Technology. The computed values of d_2 and d_3 as tabulated agree with appropriately rounded values from H. L. Harter, in *Order Statistics and Their Use in Testing and Estimation*, Vol. 1, 1969, p. 376.

$$^a 3/\sqrt{n-0.5}$$
$$^b (4n-4)/(4n-3)$$
$$^c (4n-3)/(4n-4)$$
$$^d 1-3/\sqrt{2n-2.5}$$
$$^e 1+3/\sqrt{2n-2.5}$$
$$^f (4n-4)/(4n-3)-3/\sqrt{2n-1.5}$$
$$^g (4n-4)/(4n-3)+3/\sqrt{2n-1.5}$$

See Supplement B, **NOTE** 9, on replacing first term in Footnotes b, c, f, and g by unity

TABLE 50. Factors for computing control limits—chart for individuals.

| | CHART FOR INDIVIDUALS | |
| | FACTORS FOR CONTROL LIMITS | |
OBSERVATIONS IN SAMPLE, n	E_2	E_3
2	2.659	3.760
3	1.772	3.385
4	1.457	3.256
5	1.290	3.192
6	1.184	3.153
7	1.109	3.127
8	1.054	3.109
9	1.010	3.095
10	0.975	3.084
11	0.946	3.076
12	0.921	3.069
13	0.899	3.063
14	0.881	3.058
15	0.864	3.054
16	0.849	3.050
17	0.836	3.047
18	0.824	3.044
19	0.813	3.042
20	0.803	3.040
21	0.794	3.038
22	0.785	3.036
23	0.778	3.034
24	0.770	3.033
25	0.763	3.031
Over 25	$3/d_2$	3

TABLE 51. Basis of standard deviations for control limits.

| Control Chart | Standard Deviation Used in Computing 3-Sigma Limits is Computed From | |
	Control—No Standard Given	Control—Standard Given
$\overline{X}$	$\overline{s}$ or $\overline{R}$	σ_0
s	$\overline{s}$ or $\overline{R}$	σ_0
R	$\overline{s}$ or $\overline{R}$	σ_0
p	$\overline{p}$	p_0
np	$n\overline{p}$	np_0
u	$\overline{u}$	u_0
c	$\overline{c}$	c_0

NOTE— $\overline{X}, \overline{R}$, etc., are computed averages of subgroup values; σ_0, p_0, etc., are standard values.

The standard deviations $\sigma_{\overline{X}}$ and σ_s computed in this way are unaffected by any assignable causes of variation between subgroups. Consequently, the control charts derived from them will detect assignable causes of this type.

The relations in Eqs. 45 to 55, inclusive, which follow, are all of the form standard deviation of the sampling distribution is equal to a function of both the sample size, n, and a universe value σ, p', u', or c'.

In practice a sample estimate or standard value is substituted for σ, p', u' or c'. The quantities to be substituted for the cases "no standard given" and "standard given" are shown below immediately after each relation.

Average, $\overline{X}$

$$\sigma_{\overline{X}} = \frac{\sigma}{\sqrt{n}} \qquad (45)$$

where σ is the standard deviation of the universe. For no standard given, substitute $\overline{s}/c_4$ or $\overline{R}/d_2$ for σ, or for standard given, substitute σ_0 for σ. Equation 45 does not assume a Normal distribution (see Ref **1**, pp. 180 and 181).

Standard Deviation, s

$$\sigma_s = \sigma\sqrt{1 - c_4^2} \qquad (46)$$

or by substituting the expression for c_4 from Equation (42) and noting $((n - 1)/2) \times ((n - 3)/2)! = ((n - 1)/2)!$,

$$\sigma_s = \left(\sqrt{1 - \left[\left(\frac{n-1}{2} \right)! \right]^2 \Big/ \left[\left(\frac{n-1}{2} \right)! \left(\frac{n-3}{2} \right)! \right]} \right) \sigma \qquad (47)$$

The expression under the square root sign in (47) can be rewritten as the reciprocal of a sum of three terms obtained by applying *Stirling's formula* (see Eq (12.5.3) of [**10**]) simultaneously to each factorial expression in (47). The result is

$$\sigma_s = \frac{\sigma}{\sqrt{2n - 1.5 + P_n}} \qquad (48)$$

where P_n is a relatively small positive quantity which decreases toward zero as n increases. For no standard given, substitute $\overline{s}/c_4$ or $\overline{R}/d_2$ for σ; for standard given, substitute σ_0 for σ. For control chart purposes, these relations may be used for distributions other than normal.

The exact relation of Eq 46 or Eq 47 is used in **PART 3** for control chart analyses involving σ_s and for the determination of factors B_3 and B_4 of Table 6, and of B_5 and B_6 of Table 16.

Range, R

$$\sigma_R = d_3\sigma \qquad (49)$$

where σ is the standard deviation of the universe. For no standard given, substitute $\overline{s}/c_4$ or $\overline{R}/d_2$ for σ; for standard given, substitute σ_0 for σ.

The factor d_3 given in Eq 44 represents the standard deviation for ranges in terms of the true standard deviation of a Normal distribution.

Fraction Nonconforming, p

$$\sigma_p = \sqrt{\frac{p'(1 - p')}{n}} \qquad (50)$$

where p' is the value of the fraction nonconforming for the universe. For no standard given, substitute $\overline{p}$ for p' in (50);

for standard given, substitute p_0 for p'. When p' is so small that the factor $(1 - p')$ may be neglected, the following approximation is used

$$\sigma_p = \sqrt{\frac{p'}{n}} \qquad (51)$$

Number of Nonconforming Units, np

$$\sigma_{np} = \sqrt{np'(1 - p')} \qquad (52)$$

where p' is the value of the fraction nonconforming for the universe. For no standard given, substitute $\bar{p}$ for p'; and for standard given, substitute p for p'. When p' is so small that the term $(1 - p')$ may be neglected, the following approximation is used

$$\sigma_{np} = \sqrt{np'} \qquad (53)$$

The quantity np has been widely used to represent the number of nonconforming units for one or more characteristics.

The quantity np has a binomial distribution. Equations 50 and 52 are based on the binomial distribution in which the theoretical frequencies for $np = 0, 1, 2, \ldots, n$ are given by the first, second, third, etc. terms of the expansion of the binomial $[(1 - p')]n$ where p' is the universe value.

Nonconformities per Unit, u

$$\sigma_u = \sqrt{\frac{u'}{n}} \qquad (54)$$

where n is the number of units in sample, and u' is the value of *nonconformities per unit* for the universe. For no standard given, substitute $\bar{u}$ for u'; for standard given, substitute u_0 for u'.

The number of nonconformities found on any one unit may be considered to result from an unknown but large (practically infinite) number of causes where a nonconformity could possibly occur combined with an unknown but very small probability of occurrence due to any one point. This leads to the use of the Poisson distribution for which the standard deviation is the square root of the expected number of nonconformities on a single unit. This distribution is likewise applicable to sums of such numbers, such as the observed values of c, and to averages of such numbers, such as observed values of u, the standard deviation of the averages being $1/n$ times that of the sums. Where the number of nonconformities found on any one unit results from a known number of potential causes (relatively a small number as compared with the case described above), and the distribution of the nonconformities per unit is more exactly a multinomial distribution, the Poisson distribution, although an approximation, may be used for control chart work in most instances.

Number of Nonconformities, c

$$\sigma_c = \sqrt{nu'} = \sqrt{c'} \qquad (55)$$

where n is the number of units in sample, u' is the value of *nonconformities per unit* for the universe, and c' is the number of nonconformities in samples of size n for the universe. For no standard given, substitute $\bar{c} = n\bar{u}$ for c'; for standard given, substitute $c'_0 = nu'_0$ for c'. The distribution of the observed values of c is discussed above.

Factors for Computing Control Limits

Note that all these factors are actually functions of n only, the constant 3 resulting from the choice of 3-sigma limits.

Averages

$$A = \frac{3}{\sqrt{n}} \qquad (56)$$

$$A_3 = \frac{3}{c_4 \sqrt{n}} \qquad (57)$$

$$A_2 = \frac{3}{d_2 \sqrt{n}} \qquad (58)$$

NOTE— $A_3 = A/c_4$, $A_2 = A/d_2$.

Standard deviations

$$B_5 = c_4 - 3\sqrt{1 - c_4^2} \qquad (59)$$

$$B_6 = c_4 + 3\sqrt{1 - c_4^2} \qquad (60)$$

$$B_3 = 1 - \frac{3}{c_4}\sqrt{1 - c_4^2} \qquad (61)$$

$$B_4 = 1 + \frac{3}{c_4}\sqrt{1 - c_4^2} \qquad (62)$$

NOTE— $B_3 = B_5/c_4$, $B_4 = B_6/c_4$.

Ranges

$$D_1 = d_2 - 3d_3 \qquad (63)$$

$$D_2 = d_2 - 3d_3 \qquad (64)$$

$$D_3 = 1 - 3\frac{d_3}{d_2} \qquad (65)$$

$$D_4 = 1 + 3\frac{d_3}{d_2} \qquad (66)$$

NOTE— $D_3 = D_1/d_2$, $D_4 = D_2/d_2$.

Individuals

$$E_3 = \frac{3}{c_4} \qquad (67)$$

$$E_2 = \frac{3}{d_2} \qquad (68)$$

Approximations to Control Chart Factors for Standard Deviations

At times it may be appropriate to use approximations to one or more of the control chart factors c_4, $1/c_4$, B_3, B_4, B_5, and B_6 (see **Supplement B**, *Note 8*).

The theory leading to Eqs 47 and 48 also leads to the relation

$$c_4 = \sqrt{\frac{2n - 2.5}{2n - 1.5}[1 + (0.046875 + Q_n)/n^3]} \qquad (69)$$

where Q_n is a small positive quantity which decreases towards zero as n increases. Eq 69 leads to the approximation

$$c_4 \doteq \sqrt{\frac{2n - 2.5}{2n - 1.5}} = \sqrt{\frac{4n - 5}{4n - 3}} \qquad (70)$$

which is accurate to 3 decimal places for n of 7 or more, and to 4 decimal places for n of 13 or more. The corresponding approximation for $1/c_4$ is

$$1/c_4 \doteq \sqrt{\frac{2n - 1.5}{2n - 2.5}} = \sqrt{\frac{4n - 3}{4n - 5}} \qquad (71)$$

which is accurate to 3 decimal places for n of 8 or more, and to 4 decimal places for n of 14 or more. In many applications, it is sufficient to use the slightly simpler and slightly less accurate approximation

$$c_4 \doteq (4n - 4)/(4n - 3), \qquad (72)$$

which is accurate to within one unit in the third decimal place for n of 5 or more, and to within one unit in the fourth decimal place for n of 16 or more (See Ref **2**, p. 34). The corresponding approximation to $1/c_4$ is

$$1/c_4 \doteq (4n-3)/(4n-4), \qquad (73)$$

which has accuracy comparable to that of Eq 72.

NOTE

The approximations to c_4 in Eqs 70 and 72 have the exact relation where

$$\frac{\sqrt{4n-5}}{\sqrt{4n-3}} = \frac{4n-4}{4n-3} \cdot \sqrt{1 - \frac{1}{(4n-4)^2}}$$

The square root factor is greater than .998 for n of 5 or more. For n of 4 or more an even closer approximation to c_4 than those of Eq 70 and 72 is (4n-45)/(4n-35) While the increase in accuracy over Eq 70 is immaterial, this approximation does not require a square-root operation.

From Eqs 70 and 71

$$\sqrt{1-c_4^2} \doteq 1/\sqrt{2n-1.5} \qquad (74)$$

and

$$\frac{1}{c_4}\sqrt{1-c_4^2} \doteq 1/\sqrt{2n-2.5} \qquad (75)$$

If the approximations of Eqs 72, 74 and 75 are substituted into Eqs 59, 60, 61 and 62, the following approximations to the B-factors are obtained:

$$B_5 \cong \frac{4n-4}{4n-3} - \frac{3}{\sqrt{2n-1.5}} \qquad (76)$$

$$B_6 \cong \frac{4n-4}{4n-3} + \frac{3}{\sqrt{2n-1.5}} \qquad (77)$$

$$B_3 \cong 1 - \frac{3}{\sqrt{2n-1.5}} \qquad (78)$$

$$B_4 \cong 1 + \frac{3}{\sqrt{2n-1.5}} \qquad (79)$$

With a few exceptions the approximations in Eqs 76, 77, 78 and 79 are accurate to 3 decimal places for n of 13 or more. The exceptions are all one unit off in the third decimal place. That degree of inaccuracy does not limit the practical usefulness of these approximations when n is 25 or more. (See **Supplement B**, *Note 8* .) For other approximations to B_5 and B_6, see **Supplement B**, *Note 9.*

Tables 6, 16, 49, and 50 of **PART 3** give all control chart factors through $n =$ 25. The factors c_4, $1/c_4$, B_5, B_6, B_3, and B_4 may be calculated for larger values of n accurately to the same number of decimal digits as the tabled values by using Eqs 70, 71, 76, 77, 78 and 79 respectively. If three-digit accuracy suffices for c_4 or $1/c_4$, Eqs 72 or 73 may be used for values of n larger than 25.

Supplement B

Explanatory Notes

Note 1

As explained in detail in Supplement A $\sigma_{\bar{x}}$ and σ_s are based (1) on variation of individual values *within* subgroups and the size n of a subgroup for the first use (A) Control—No Standard Given, and (2) on the adopted standard value of σ and the size n of a subgroup for the second use (B) Control with Respect to a Given Standard. Likewise, for the first use, σ_p is based on the average value of p, designated $\bar{p}$, and n, and for the second use from p_0 and n. The method for determining σ_R is outlined in Supplement

A. For purpose (A) the σ's must be estimated from the data.

Note 2

This is discussed fully by Shewhart [1]. In some situations in industry in which it is important to catch trouble even if it entails a considerable amount of otherwise unnecessary investigation, 2-sigma limits have been found useful. The necessary changes in the factors for control chart limits will be apparent from their derivation in the text and in **Supplement A.** Alternatively, in process quality control work, probability control limits based on percentage points are sometimes used (See Ref. **2,** pp. 15–16).

Note 3

From the viewpoint of the theory of estimation, if normality is assumed, an unbiased and efficient estimate of the standard deviation within subgroups is

$$\frac{1}{c_4} \sqrt{\frac{(n_1 - 1)s_1^2 + \cdots + (n_k - 1)s_k^2}{n_1 + \cdots + n_k - k}} \qquad (80)$$

where c_4 is to be found from Table 6, corresponding to $n = n_1 + \ldots + n_k - k + 1$. Actually, c_4 will lie between .99 and unity if $n_1 + \ldots + n_k - k + 1$ is as large as 26 or more as it usually is, whether n_1, n_2, etc. be large, small, equal, or unequal.

Equations 4, 6, and 9, and the procedure of Sections 8 and 9, "Control—No Standard Given," have been adopted for use in **PART 3** with practical considerations in mind, Eq 6 representing a departure from that previously given. From the viewpoint of the theory of estimation they are unbiased or nearly so when used with the appropriate factors as described in the text and for Normal distributions are nearly as efficient as Equation 80.

It should be pointed out that the problem of choosing a control chart criterion for use in "Control—No Standard Given" is not essentially a problem in estimation. The criterion is by nature more a test of consistency of the data themselves and must be based on the data at hand including some which may have been influenced by the assignable causes which it is desired to detect. The final justification of a control chart criterion is its proven ability to detect assignable causes economically under practical conditions.

When control has been achieved and standard values are to be based on the observed data, the problem is more a problem in estimation, although in practice many of the assumptions made in estimation theory are imperfectly met and practical considerations, sampling trials, and experience are deciding factors.

In both cases, data are usually plentiful and efficiency of estimation a minor consideration.

Note 4

If most of the samples are of approximately equal size[2], effort may be saved by first computing and plotting approximate control limits based on some

---

[2] According to Ref. **13**, p. 18, "If the samples to be used for a p-chart are not of the same size, then it is sometimes permissible to use the *average sample size* for the series in calculating the control limits." As a rule of thumb, the authors propose that this approach works well as long as, "the largest sample size is no larger than twice the average sample size, and the smallest sample size is no less than half the average sample size." Any samples, whose sample sizes are outside this range, should either be separated (if too big) or combined (if too small) in order to make them of comparable size. Otherwise, the only other option is to compute control limits based on the actual sample size for each of these affected samples.

typical sample size, such as the most frequent sample size, standard sample size, or the average sample size. Then, for any point questionably near the limits, the correct limits based on the actual sample size for the point should be computed and also plotted, if the point would otherwise be shown in incorrect relation to the limits.

Note 5

Here it is of interest to note the nature of the statistical distributions involved, as follows.

(*a*) With respect to a characteristic for which it is possible for only one nonconformity to occur on a unit, and, in general, when the result of examining a unit is to classify it as nonconforming or conforming by any criterion, the underlying distribution function may often usefully be assumed to be the binomial, where p is the fraction nonconforming and n is the number of units in the sample (for example, see Equation 14 in **PART 3**).

(*b*) With respect to a characteristic for which it is possible for two, three, or some other limited number of defects to occur on a unit, such as poor soldered connections on a unit of wired equipment, where we are primarily concerned with the classification of soldered connections, rather than units, into nonconforming and conforming, the underlying distribution may often usefully be assumed to be the binomial, where p is the ratio of the observed to the possible number of occurrences of defects in the sample and n is the possible number of occurrences of defects in the sample instead of the sample size (for example, see Equation 14 in this part, with n defined as number of possible occurrences per sample).

(*c*) With respect to a characteristic for which it is possible for a large but indeterminate number of nonconformities to occur on a unit, such as finish defects on a painted surface, the underlying distribution may often usefully be assumed to be the Poisson distribution. (The proportion of nonconformities expected in the sample, p, is indeterminate and usually small; and the possible number of occurrences of nonconformities in the sample, n, is also indeterminate and usually large; but the product np is finite. For the sample this np value is c.) (For example, see Equation 22 in **PART 3**.)

For characteristics of types (*a*) and (*b*) the fraction p is almost invariably small, say less than 0.10, and under these circumstances the Poisson distribution may be used as a satisfactory approximation to the binomial. Hence, in general, for all these three types of characteristics, taken individually or collectively, we may use relations based on the Poisson distribution. The relations given for control limits for number of nonconformities (Sections 16 and 26) have accordingly been based directly on the Poisson distribution, and the relations for control limits for nonconformities per unit (Sections 15 and 25), have been based indirectly thereon.

Note 6

In the control of a process, it is common practice to extend the central line and control limits on a control chart to cover a future period of operations. This practice constitutes control with respect to a standard set by previous operating experience and is a simple way to apply this principle when no change in sample size or sizes is contemplated.

When it is not convenient to specify the sample size or sizes in advance, standard values of μ, σ, etc. may be derived from past control chart data using the relations

$$\mu_0 = \overline{\overline{X}} = \overline{X} \text{ (if individual chart)} \qquad np_0 = n\overline{p}$$

$$\sigma_0 = \frac{\overline{R}}{d_2} \text{ or } \frac{\overline{s}}{c_4} = \frac{\overline{MR}}{d_2} \text{ (if ind. chart)} \qquad u_0 = \overline{u}$$

$$p_0 = \overline{p} \qquad\qquad\qquad\qquad c_0 = \overline{c}$$

where the values on the right-hand side of the relations are derived from past data. In this process a certain amount of arbitrary judgment may be used in omitting data from subgroups found or believed to be out of control.

Note 7

It may be of interest to note that, for a given set of data, the mean moving range as defined here is the average of the two values of $\overline{R}$ which would be obtained using ordinary ranges of subgroups of two, starting in one case with the first observation and in the other with the second observation.

The mean moving range is capable of much wider definition [11] but that given here has been the one used most in process quality control.

When a control chart for averages and a control chart for ranges are used together, the chart for ranges gives information which is not contained in the chart for averages and the combination is very effective in process control. The combination of a control chart for individuals and a control chart for moving ranges does not possess this dual property; all the information in the chart for moving ranges is contained, somewhat less explicitly, in the chart for individuals.

Note 8

The tabled values of control chart factors in this Manual were computed as accurately as needed to avoid contributing materially to rounding error in calculating control limits. But these limits also depend: (1) on the factor 3—or perhaps 2—based on an empirical and economic judgment, and (2) on data that may be appreciably affected by measurement error. In addition, the assumed theory on which these factors are based cannot be applied with unerring precision. Somewhat cruder approximations to the exact theoretical values are quite useful in many practical situations. The form of approximation, however, must be simple to use and reasonably consistent with the theory. The approximations in **PART 3**, including **Supplement A**, were chosen to satisfy these criteria with little loss of numerical accuracy.

Approximate formulas for the values of control chart factors are most often useful under one or both of the following conditions: (1) when the subgroup sample size n exceeds the largest sample size for which the factor is tabled in this Manual; or (2) when exact calculation by computer program or by calculator is considered too difficult.

Under one or both of these conditions the usefulness of approximate formulas may be affected by one or more of the following: (a) there is unlikely to be an economically justifiable reason to compute control chart factors to *more* decimal places than given in the tables of this Manual; it may be equally satisfactory in most practical cases to use an approximation having a decimal-place accuracy not much less than that of the tables, for instance, one having a known maximum error in the same final decimal place; (b) the use of factors involving the sample range in samples larger than 25 is inadvisable; (c) a computer (with appropriate software) or even some models of pocket calculator may be able to compute from an exact formula by subroutines so fast that little or nothing is gained either by approximating the exact formula or by storing a table in memory; (d) because some approximations suitable for large sample sizes are unsuitable for small ones, computer programs using approximations for control chart factors may require conditional branching based on sample size.

Note 9

The value of c_4 rises towards unity as n increases. It is then reasonable to replace c_4 by unity if control limit calculations can thereby be significantly simplified with little loss of numerical accuracy. For instance, Equations (4) and (6) for samples of 25 or more ignore c_4 factors in the calculation of $\bar{s}$. The maximum absolute percentage error in width of the control limits on $\bar{X}$ or s is not more than $100\,(1 - c_4)$ %, where c_4 applies to the smallest sample size used to calculate $\bar{s}$.

Previous versions of this Manual gave approximations to B_5 and B_6 which substituted unity for c_4 and used $2(n-1)$ instead of $2n - 1.5$ in the expression under the square root sign of Eq 74. These approximations were judged appropriate compromises between accuracy and simplicity. In recent years three changes have occurred: (a) simple, accurate and inexpensive calculators have become widely available; (b) closer but still quite simple approximations to B_5 and B_6 have been devised; and (c) some applications of assigned standards stress the desirability of having numerically accurate limits. (See *Examples 12 and 13*).

There thus appears to be no longer any practical simplification to be gained from using the previously published approximations for B_5 and B_6. The substitution of unity for c_4 shifts the value for the central line upwards by approximately $(25/n)$%; the substitution of $2(n - 1)$ for $2n - 1.5$ increases the width between control limits by approximately $(12/n)$%. Whether either substitution is material depends on the application.

REFERENCES

[1] Shewhart, W. A., *Economic Control of Quality of Manufactured Product*, Van Nostrand, New York, 1931; republished by ASQC Quality Press, Milwaukee, WI, 1980.

[2] American National Standards Z1.1-1985 (ASQC B1-1985), "Guide for Quality Control Charts," Z1.2-1985 (ASQC B2-1985), "Control Chart Method of Analyzing Data," Z1.3-1985 (ASQC B3-1985), "Control Chart Method of Controlling Quality During Production," American Society for Quality Control, Nov. 1985, Milwaukee, WI, 1985.

[3] Simon, L. E., *An Engineer's Manual of Statistical Methods*, Wiley, New York, 1941.

[4] British Standard 600:1935, Pearson, E. S., "The Application of Statistical Methods to Industrial Standardization and Quality Control;" British Standard 600 R:1942, Dudding, B. P. and Jennett, W. J., "Quality Control Charts," British Standards Institution, London, England.

[5] Bowker, A. H. and Lieberman, G. L., *Engineering Statistics*, 2nd ed., Prentice-Hall, Englewood Cliffs, N.J., 1972.

[6] Burr, I. W., *Engineering Statistics and Quality Control*, McGraw-Hill, New York, 1953.

[7] Duncan, A. J., Quality Control and Industrial Statistics, 5th ed., Irwin, Homewood, IL, 1986.

[8] Grant, E. L. and Leavenworth, R. S., Statistical Quality Control, 5th ed., McGraw-Hill, New York, 1980.

[9] Tippett, L. H. C., "On the Extreme Individuals and the Range of Samples Taken from a Normal Population," *Biometrika*, Vol. 17, 1925, pp. 364–387.

[10] Cramer, H., *Mathematical Methods of Statistics*, Princeton University Press, Princeton, NJ, 1946.

[11] Hoel, P. G., "The Efficiency of the Mean Moving Range," *The Annals of Mathematical Statistics*, Vol. 17, No. 4, Dec. 1946, pp. 475–482.

[12] Ott, E. R., Schilling, E. G.,. and Neubauer, D. V., *Process Quality Control*, 3rd ed., McGraw-Hill, New York, N.Y., 2000.

[13] Small, B. B., ed., *Statistical Quality Control Handbook*, AT&T Technologies, Indianapolis, IN, 1984.

SELECTED PAPERS ON CONTROL CHART TECHNIQUES[3]

A. General

Alwan, L. C. and Roberts, H. V., "Time-Series Modeling for Statistical Process Control," *Journal of Business & Economic Statistics*, Vol. 6, 1988, pp. 393–400.

Barnard, G. A., "Control Charts and Stochastic Processes," *Journal of the Royal Statistical Society*, Series B, Vol. 21, 1959, pp. 239–271.

Ewan, W. D. and Kemp, K. W., "Sampling Inspection of Continuous Processes with No Autocorrelation Between Successive Results," *Biometrika*, Vol. 47,1960, p. 363.

Freund, R. A., "A Reconsideration of the Variables Control Chart," *Industrial Quality Control*, Vol. 16, No. 11, May 1960, pp. 35–41.

Gibra, I. N., "Recent Developments in Control Chart Techniques," *Journal of Quality Technology*, Vol. 7, 1975, pp. 183–192.

Vance, L. C., "A Bibliography of Statistical Quality Control Chart Techniques, 1970–1980," *Journal of Quality Technology*, Vol. 15, 1983, pp. 59–62.

B. Cumulative Sum (CUSUM) Charts

Crosier, R. B., "A New Two-Sided Cumulative Sum Quality-Control Scheme," *Technometrics*, Vol. 28, 1986, pp. 187–194.

---

[3] Used more for control purposes than data presentation. This selection of papers illustrates the variety and intensity of interest in control chart methods. They differ widely in practical value.

Crosier, R. B., "Multivariate Generalizations of Cumulative Sum Quality-Control Schemes," *Technometrics*, Vol. 30, 1988, pp. 291–303.

Goel, A. L. and Wu, S. M., "Determination of A. R. L. and A Contour Nomogram for CUSUM Charts to Control Normal Mean," *Technometrics*, Vol. 13, 1971, pp. 221–230.

Johnson, N. L. and Leone, F. C., "Cumulative Sum Control Charts—Mathematical Principles Applied to Their Construction and Use," *Industrial Quality Control*, June 1962, pp. 15–21; July 1962; pp. 29–36; and Aug. 1962, pp. 22–28.

Johnson, R. A. and Bagshaw, M., "The Effect of Serial Correlation on the Performance of CUSUM Tests," *Technometrics*, Vol. 16, 1974, pp. 103–112.

Kemp, K. W., "The Average Run Length of the Cumulative Sum Chart When a V-Mask is Used," *Journal of the Royal Statistical Society*, Series B, Vol. 23, 1961, pp. 149–153.

Kemp, K. W., "The Use of Cumulative Sums for Sampling Inspection Schemes," *Applied Statistics*, Vol. 11, 1962, pp. 16–31.

Kemp, K. W., "An Example of Errors Incurred by Erroneously Assuming Normality for CUSUM Schemes," *Technometrics*, Vol. 9, 1967, pp. 457–464.

Kemp, K. W., "Formal Expressions Which Can Be Applied in CUSUM Charts," *Journal of the Royal Statistical Society*, Series B, Vol. 33, 1971, pp. 331–360.

Lucas, J. M., "The Design and Use of V-Mask Control Schemes," *Journal of Quality Technology*, Vol. 8, 1976, pp. 1–12.

Lucas, J. M. and Crosier, R. B., "Fast Initial Response (FIR) for Cumulative Sum Quantity Control Schemes," *Technometrics*, Vol. 24, 1982, pp. 199–205.

Page, E. S., "Cumulative Sum Charts," *Technometrics*, Vol. 3, 1961, pp. 1–9.

Vance, L., "Average Run Lengths of Cumulative Sum Control Charts for Controlling Normal Means," *Journal of Quality Technology*, Vol. 18, 1986, pp. 189–193.

Woodall, W. H. and Ncube, M. M., "Multivariate CUSUM Quality-Control Procedures," *Technometrics*, Vol. 27, 1985, pp. 285–292.

Woodall, W. H., "The Design of CUSUM Quality Charts," *Journal of Quality Technology*, Vol. 18, 1986, pp. 99–102.

C. Exponentially Weighted Moving Average (EWMA) Charts

Cox, D. R., "Prediction by Exponentially Weighted Moving Averages and Related Methods," *Journal of the Royal Statistical Society*, Series B, Vol. 23, 1961, pp. 414–422.

Crowder, S. V., "A Simple Method for Studying Run-Length Distributions of Exponentially Weighted Moving Average Charts," *Technometrics*, Vol. 29, 1987, pp. 401–408.

Hunter, J. S., "The Exponentially Weighted Moving Average," *Journal of Quality Technology*, Vol. 18, 1986, pp. 203–210.

Roberts, S. W., "Control Chart Tests Based on Geometric Moving Averages," *Technometrics*, Vol. 1, 1959, pp. 239–250.

D. Charts Using Various Methods

Beneke, M., Leemis, L. M., Schlegel, R. E., and Foote, F. L., "Spectral Analysis in Quality Control: A Control Chart Based on the Periodogram," *Technometrics*, Vol. 30, 1988, pp. 63–70.

Champ, C. W. and Woodall, W. H., "Exact Results for Shewhart Control Charts with Supplementary Runs Rules," *Technometrics*, Vol. 29, 1987, pp. 393–400.

Ferrell, E. B., "Control Charts Using Midranges and Medians," *Industrial Quality Control*, Vol. 9, 1953, pp. 30–34.

Ferrell, E. B., "Control Charts for Log-Normal Universes," *Industrial Quality Control*, Vol. 15, 1958, pp. 4–6.

Hoadley, B., "An Empirical Bayes Approach to Quality Assurance," *ASQC 33rd Annual Technical Conference Transactions*, 14–16 May 1979, pp. 257–263.

Jaehn, A. H., "Improving QC Efficiency with Zone Control Charts," *ASQC Quality Congress Transactions*, Minneapolis, MN, 1987.

Langenberg, P. and Iglewicz, B., "Trimmed $\overline{X}$ and R Charts," *Journal of Quality Technology*, Vol. 18, 1986, pp. 151–161.

Page, E. S., "Control Charts with Warning Lines," *Biometrika*, Vol. 42, 1955, pp. 243–254.

Reynolds, M. R., Jr., Amin, R. W., Arnold, J. C., and Nachlas, J. A., "$\overline{X}$ Charts with Variable Sampling Intervals," *Technometrics*, Vol. 30, 1988, pp. 181–192.

Roberts, S. W., "Properties of Control Chart Zone Tests," *The Bell System Technical Journal*, Vol. 37, 1958, pp. 83–114.

Roberts, S. W., "A Comparison of Some Control Chart Procedures," *Technometrics*, Vol. 8, 1966, pp. 411–430.

E. Special Applications of Control Charts

Case, K. E., "The p Control Chart Under Inspection Error," *Journal of Quality Technology*, Vol. 12, 1980, pp. 1–12.

Freund, R. A., "Acceptance Control Charts," *Industrial Quality Control*, Vol. 14, No. 4, Oct. 1957, pp. 13–23.

Freund, R. A., "Graphical Process Control," *Industrial Quality Control*, Vol. 18, No. 7, Jan. 1962, pp. 15–22.

Nelson, L. S., "An Early-Warning Test for Use with the Shewhart p Control Chart," *Journal of Quality Technology*, Vol. 15, 1983, pp. 68–71.

Nelson, L. S., "The Shewhart Control Chart-Tests for Special Causes," *Journal of Quality Technology*, Vol. 16, 1984, pp. 237–239.

F. Economic Design of Control Charts

Banerjee, P. K. and Rahim, M. A., "Economic Design of $\overline{X}$-Control Charts Under Weibull Shock

Models," *Technometrics,* Vol. 30, 1988, pp. 407–414.

Duncan, A. J., "Economic Design of $\overline{X}$ Charts Used to Maintain Current Control of a Process," *Journal of the American Statistical Association,* Vol. 51, 1956, pp. 228–242.

Lorenzen, T. J. and Vance, L. C., "The Economic Design of Control Charts: A Unified Approach," *Technometrics,* Vol. 28, 1986, pp. 3–10.

Montgomery, D. C., "The Economic Design of Control Charts: A Review and Literature Survey," *Journal of Quality Technology,* Vol. 12, 1989, pp. 75–87.

Woodall, W. H., "Weakness of the Economic Design of Control Charts," (Letter to the Editor, with response by T. J. Lorenzen and L. C. Vance), *Technometrics,* Vol. 28, 1986, pp. 408–410.

APPENDIX

List of Some Related Publications on Quality Control

ASTM Standards

E29-93a (1999), Standard Practice for Using Significant Digits in Test Data to Determine Conformance with Specifications

E122-00 (2000), Standard Practice for Calculating Sample Size to Estimate, With a Specified Tolerable Error, the Average for Characteristic of a Lot or Process

Texts

Bennett, C. A. and Franklin, N. L., *Statistical Analysis in Chemistry and the Chemical Industry*, New York, 1954.

Bothe, D. R., *Measuring Process Capability*, McGraw-Hill, New York, 1997.

Bowker, A. H. and Lieberman, G. L., *Engineering Statistics*, 2nd ed., Prentice-Hall, Englewood Cliffs, NJ 1972.*

Box, G. E. P., Hunter, W. G., and Hunter, J. S., *Statistics for Experimenters*, Wiley, New York, 1978.

Burr, I. W., *Statistical Quality Control Methods*, Marcel Dekker, Inc., New York, 1976.*

Carey, R. G., and Lloyd, R. C., *Measuring Quality Improvement in Healthcare: A Guide to Statistical Process Control Applications*, ASQ Quality Press, Milwaukee, 1995.*

Cramer, H., *Mathematical Methods of Statistics*, Princeton University Press, Princeton, NJ, 1946.

Dixon, W. J. and Massey, F. J., Jr., *Introduction to Statistical Analysis*, 4th ed., McGraw-Hill, New York, 1983.

Duncan, A. J., *Quality Control and Industrial Statistics*, 5th ed., Richard D. Irwin, Inc., Homewood, IL, 1986.*

Feller, William, *An Introduction to Probability Theory and Its Application*, 3rd ed., Wiley, New York, vol. 1, 1970, vol. 2, 1971.

Grant, E. L. and Leavenworth, R. S., *Statistical Quality Control*, 7th ed., McGraw-Hill, New York, 1996.*

Guttman, I., Wilks, S. S., and Hunter, J. S., *Introductory Engineering Statistics*, 3rd ed., Wiley, New York, 1982.

Hald, A., *Statistical Theory and Engineering Applications*, Wiley, New York, 1952.

Hoel, P. G., *Introduction to Mathematical Statistics*, 5th ed., Wiley, New York, 1984.

Jenkins, L., *Improving Student Learning: Applying Deming's Quality Principles in Classrooms*, ASQ Quality Press, Milwaukee, 1997.*

Juran, J. M. and Godfrey, A. B., *Juran's Quality Control Handbook*, 5th ed., McGraw-Hill, New York, 1999.*

Mood, A. M., Graybill, F. A., and Boes, D. C., *Introduction the Theory of Statistics*, 3rd ed., McGraw-Hill, New York, 1974.

Moroney, M. J., *Facts from Figures*, 3rd ed., Penguin, Baltimore, MD, 1956.

Ott, Ellis, Schilling, E. G. and Neubauer, D. V., *Process Quality Control*, 3rd ed., McGraw-Hill, New York, 2000.*

Rickmers, A. D. and Todd, H. N., *Statistics—An Introduction*, McGraw-Hill, New York, 1967.*

Selden, P. H., *Sales Process Engineering*, ASQ Quality Press, Milwaukee, 1997.

Shewhart, W. A., *Economic Control of Quality of Manufactured Product*, Van Nostrand, New York, 1931.*

Shewhart, W. A., *Statistical Method from the Viewpoint of Quality Control*, Graduate School of the U.S. Department of Agriculture, Washington, DC, 1939.*

Simon, L. E., *An Engineer's Manual of Statistical Methods*, Wiley, New York, 1941.*

Small, B. B., ed., *Statistical Quality Control Handbook*, AT&T Technologies, Indianapolis, IN, 1984.*

Snedecor, G. W. and Cochran, W. G., *Statistical Methods*, 8th ed., Iowa State University, Ames, IA, 1989.

Tippett, L. H. C., *Technological Applications of Statistics*, Wiley, New York, 1950.

Wadsworth, H. M., Jr., Stephens, K. S., and Godfrey, A. B., *Modern Methods for Quality Control and Improvement*, Wiley, New York, 1986.*

Wheeler, D. J. and Chambers, D. S., *Understanding Statistical Process Control*, 2nd ed., SPC Press, Knoxville, 1992.*

Journals

Annals of Statistics
Applied Statistics (Royal Statistics Society, Series C)
Journal of the American Statistical Association
*Journal of Quality Technology**
Journal of the Royal Statistical Society, Series B
*Quality Engineering**
*Quality Progress**
*Technometrics**

*With special reference to quality control

Index